Tube and Sheet Metal Forming Processes and Applications

Tube and Sheet Metal Forming Processes and Applications

Editors

Gabriel Centeno Báez
Maria Beatriz Silva

MDPI • Basel • Beijing • Wuhan • Barcelona • Belgrade • Manchester • Tokyo • Cluj • Tianjin

Editors

Gabriel Centeno Báez
Mechanical and
Manufacturing Engineering
University of Seville
Seville
Spain

Maria Beatriz Silva
IDMEC
Instituto Superior Tecnico -
Universidade de Lisboa
Lisbon
Portugal

Editorial Office
MDPI
St. Alban-Anlage 66
4052 Basel, Switzerland

This is a reprint of articles from the Special Issue published online in the open access journal *Metals* (ISSN 2075-4701) (available at: www.mdpi.com/journal/metals/special_issues/tube_sheet_forming).

For citation purposes, cite each article independently as indicated on the article page online and as indicated below:

LastName, A.A.; LastName, B.B.; LastName, C.C. Article Title. *Journal Name* **Year**, *Volume Number*, Page Range.

ISBN 978-3-0365-4334-5 (Hbk)
ISBN 978-3-0365-4333-8 (PDF)

Contents

Preface to "Tube and Sheet Metal Forming Processes and Applications"

At present, the manufacturing industry is focused on the production of lighter weight components with better mechanical properties and always fulfilling all the environmental requirements. These challenges have caused a need for developing manufacturing processes in general, including obviously those devoted in particular to the development of thin-walled metallic shapes, as is the case with tubular and sheet metal parts and devices.

This Special Issue is thus devoted to research in the fields of sheet metal forming and tube forming, and their applications, including both experimental and numerical approaches and using a variety of scientific and technological tools, such as forming limit diagrams (FLDs), analysis on formability and failure, strain analysis based on circle grids or digital image correlation (DIC), and finite element analysis (FEA), among others.

In this context, we are pleased to present this Special Issue dealing with recent studies in the field of tube and sheet metal forming processes and their main applications within different high-tech industries, such as the aerospace, automotive, or medical sectors, among others.

Gabriel Centeno Báez and Maria Beatriz Silva
Editors

Editorial

Tube and Sheet Metal Forming Processes and Applications

Gabriel Centeno [1,*] and Maria Beatriz Silva [2]

[1] Department of Mechanical and Manufacturing Engineering, School of Engineering, University of Seville, Camino de los Descubrimientos s/n, 41092 Seville, Spain
[2] IDMEC, Instituto Superior Técnico, Universidade de Lisboa, Av. Rovisco Pais, 1049-001 Lisboa, Portugal; beatriz.silva@tecnico.ulisboa.pt
* Correspondence: gaceba@us.es; Tel.: +34-954-485-965

Citation: Centeno, G.; Silva, M.B. Tube and Sheet Metal Forming Processes and Applications. *Metals* **2022**, *12*, 553. https://doi.org/10.3390/met12040553

Received: 21 March 2022
Accepted: 23 March 2022
Published: 25 March 2022

Publisher's Note: MDPI stays neutral with regard to jurisdictional claims in published maps and institutional affiliations.

1. Introduction

In the late 1960s, pioneer works by Keeler [1] and Goodwin [2] established the initial procedures for characterizing metal sheet formability based on the use of circle grid analysis (CGA) techniques, allowing for the determination of the in-plane strains on the surface of sheet metal formed parts. Later, in the early 1980s, Embury and Duncan [3] introduced what they called 'formability maps', currently known as forming limit diagrams (FLDs) [4], allowing for the plotting of the values of the critical strains at the onset of failure, along with the strain distribution attained at the forming process of a certain industrial part or component. These research works allowed the creation of the current framework for the analysis of sheet metal forming, also extensible to tube forming.

On the other hand, the current manufacturing industry focuses on the production of light-weight components with better mechanical properties, always fulfilling the increasingly more strict environmental requirements. These challenges have resulted in the requirement for the development of manufacturing processes in general, including, evidently, those devoted in particular to the development of thin-walled metallic shapes, as is the case with tubular and sheet metal parts and devices.

Thus, this Special Issue is devoted to research work in the field of sheet metal forming, tube forming, and their applications, including both experimental and numerical approaches and using a variety of scientific and technological tools, such as the above-mentioned FLDs, analysis on formability and failure, strain analysis based on circle grids or digital image correlation (DIC), and finite element analysis (FEA), among others.

The contributions presented in this Special Issue are discussed in the following section, and were originally invited to deal with recent studies in the field of tube and sheet metal forming processes and their main applications within different high-tech industries, such as the aerospace, automotive and medical sectors, among others.

2. Contributions

These topics were addressed in several high-quality scientific papers within this Special Issue. In what follows, the contents of the published manuscripts are briefly summarized.

Some of these contributions focused on material plastic behavior, as is the case in the work by Fang et al. [5], focusing on the direct assessment of the R-value in sheet metal based on the use of multicamera DIC systems, or the analysis of strain-hardening viscoplastic wide sheets submitted to bending under tension by Alexandrov and Lyamina [6]. Additionally, in this regard, the paper by Shahzamanian et al. [7] presented a numerical study of the influence of superimposed hydrostatic pressure on the damage mechanism by shear in sheet metal forming through the use of the shear modified GTN model to understand the effect of pressure on the shear damage mechanism.

Incremental sheet forming (ISF) was another topic of relevance in this Special Issue, dealt with in the work by Bautista-Monsalve et al. [8] through a novel machine-learning-based procedure for determining the surface finish quality of parts obtained by heat-assisted

SPIF. Additionally, the work by Suntaxi et al. [9] dealt with ISF, although in this case, concerning the multistage SPIF of thin-walled tubes from a numerical perspective. Other papers analyzing tube forming were carried out by Standley and Knezevic [10] dealing with the manufacturing of ultrafine metallic tubular structures by accumulative extrusion bonding, or the paper by Kishimoto et al. [11] which analyzed the deformation behavior causing the excessive thinning of micro metal tubes in hollow sinking.

Other contributions were dedicated to technological applications, such as the medical field in the case of Palumbo et al. [12], proposing an approach for the manufacture of cranial prostheses in sheet metal forming, the use of additive manufacturing by Tondini et al. [13] for the manufacturing of polymer tools for use in sheet metal forming, or the work by Hoffmann et al. [14] studying the reduction in warping in kinematic L-profile bending using local heating.

This compilation of research works has generously contributed to the success of this very interesting and high-quality Special Issue of *Metals*, devoted to "Tube and Sheet Metal Forming Processes and Applications".

Funding: The authors would like to express their gratitude for the funding received through grant numbers US-1263138 US/JUNTA/FEDER_UE within "Proyectos I + D + i FEDER Andalucía 2014–2020" and P18-RT-3866 US/JUNTA/FEDER_UE funded under "PAIDI 2020: Proyectos I + D + i", as well as the Fundação para a Ciência e da Tecnologia of Portugal, through IDMEC under LAETA, project UIDB/50022/2020.

Acknowledgments: The guest editors would like to especially acknowledge all of the authors who contributed their excellent work to this Special Issue. Furthermore, we would also like to thank all the reviewers for their outstanding reviews of the manuscripts submitted and for providing help-ful and constructive comments. The guest editors would like to thank the Editorial Office involved in the preparation, editing, and management of this Special Issue. We would not have been able to reach the final collection of high-quality papers without the joint efforts.

Conflicts of Interest: The authors declare no conflict of interest.

References

1. Keeler, S.P. *Circular Grid System-A Valuable Aid for Evaluating Sheet Metal Formability*; SAE Technical Papers; SAE International: Warrendale, PA, USA, 1968.
2. Goodwin, G.M. *Application of Strain Analysis to Sheet Metal Forming Problems in the Press Shop*; SAE Technical Papers; SAE International: Warrendale, PA, USA, 1968.
3. Embury, J.D.; Duncan, J.L. Formability Maps. *Annu. Rev. Mater. Sci.* **1981**, *11*, 505–521. [CrossRef]
4. Centeno, G.; Martínez-Donaire, A.J.; Morales-Palma, D.; Vallellano, C.; Silva, M.B.; Martins, P.A.F. Novel experimental techniques for the determination of the forming limits at necking and fracture. *Mater. Form. Mach. Res. Dev.* **2015**, *2*, 1–24. [CrossRef]
5. Fang, S.; Zheng, X.; Zheng, G.; Zhang, B.; Guo, B.; Yang, L. A new and direct r-value measurement method of sheet metal based on multi-camera dic system. *Metals* **2021**, *11*, 1401. [CrossRef]
6. Alexandrov, S.; Lyamina, E. Analysis of Strain-Hardening Viscoplastic Wide Sheets Subject to Bending under Tension. *Metals* **2022**, *12*, 118. [CrossRef]
7. Shahzamanian, M.; Thomsen, C.; Partovi, A.; Xu, Z.; Wu, P. Numerical study about the influence of superimposed hydrostatic pressure on shear damage mechanism in sheet metals. *Metals* **2021**, *11*, 1193. [CrossRef]
8. Bautista-Monsalve, F.; García-Sevilla, F.; Miguel, V.; Naranjo, J.; Manjabacas, M.C. A novel machine-learning-based procedure to determine the surface finish quality of titanium alloy parts obtained by heat assisted single point incremental forming. *Metals* **2021**, *11*, 1287. [CrossRef]
9. Suntaxi, C.; Centeno, G.; Silva, M.B.; Vallellano, C.; Martins, P.A.F. Tube expansion by single point incremental forming: An experimental and numerical investigation. *Metals* **2021**, *11*, 1481. [CrossRef]
10. Standley, M.R.; Knezevic, M. Towards manufacturing of ultrafine-laminated structures in metallic tubes by accumulative extrusion bonding. *Metals* **2021**, *11*, 389. [CrossRef]
11. Kishimoto, T.; Sakaguchi, H.; Suematsu, S.; Tashima, K.; Kajino, S.; Gondo, S.; Suzuki, S. Deformation behavior causing excessive thinning of outer diameter of micro metal tubes in hollow sinking. *Metals* **2020**, *10*, 1315. [CrossRef]
12. Palumbo, G.; Ambrogio, G.; Crovace, A.; Piccininni, A.; Cusanno, A.; Guglielmi, P.; De Napoli, L.; Serratore, G. A Structured Approach for the Design and Manufacturing of Titanium Cranial Prostheses via Sheet Metal Forming. *Metals* **2022**, *12*, 293. [CrossRef]

13. Tondini, F.; Basso, A.; Arinbjarnar, U.; Nielsen, C.V. The performance of 3d printed polymer tools in sheet metal forming. *Metals* **2021**, *11*, 1256. [CrossRef]
14. Hoffmann, E.; Meya, R.; Tekkaya, A.E. Reduction of warping in kinematic l-profile bending using local heating. *Metals* **2021**, *11*, 1146. [CrossRef]

Article

A Structured Approach for the Design and Manufacturing of Titanium Cranial Prostheses via Sheet Metal Forming

Gianfranco Palumbo [1,*], Giuseppina Ambrogio [2], Alberto Crovace [3], Antonio Piccininni [1], Angela Cusanno [1], Pasquale Guglielmi [1], Luigi De Napoli [2] and Giuseppe Serratore [2]

1 Department of Mechanics, Mathematics and Management, Polytechnic University of Bari, Via Orabona 4, 70125 Bari, Italy; antonio.piccininni@poliba.it (A.P.); angela.cusanno@poliba.it (A.C.); pasquale.guglielmi@poliba.it (P.G.)

2 Department of Mechanical, Energy and Management Engineering, University of Calabria, 87036 Rende, Italy; giuseppina.ambrogio@unical.it (G.A.); luigi.denapoli@unical.it (L.D.N.); giuseppe.serratore@unical.it (G.S.)

3 Department of Basic Medical Sciences, Neuroscience and Sense Organs, University of Bari Aldo Moro, Piazza Umberto I 1, 70121 Bari, Italy; alberto.crovace@uniba.it

* Correspondence: gianfranco.palumbo@poliba.it

Abstract: Currently, the growing need for highly customized implants has become one of the key aspects to increase the life expectancy and reduce time and costs for prolonged hospitalizations due to premature failures of implanted prostheses. According to the literature, several technological solutions are considered suitable to achieve the necessary geometrical complexity, from the conventional subtractive approaches to the more innovative additive solutions. In the case of cranial prostheses, which must guarantee a very good fitting of the region surrounding the implant in order to minimize micromotions and reduce infections, the need of a product characterized by high geometrical complexity combined with both strength and limited weight, has pushed the research towards the adoption of manufacturing processes able to improve the product's quality but being fast and flexible enough. The attention has been thus focused in this paper on sheet metal forming processes and, namely on the Single Point Incremental Forming (SPIF) and the Superplastic Forming (SPF). In particular, the complete procedure to design and produce titanium cranial prostheses for in vivo tests is described: starting from Digital Imaging and COmmunications in Medicine (DICOM) images of the ovine animal, the design was conducted and the production process simulated to evaluate the process parameters and the production set up. The forming characteristics of the prostheses were finally evaluated in terms of thickness distributions and part's geometry. The effectiveness of the proposed methodology has been finally assessed through the implantation of the manufactured prostheses in sheep.

Keywords: Ti-6Al-4V ELI; superplastic forming; single point incremental forming; custom prosthesis; in vivo tests

Citation: Palumbo, G.; Ambrogio, G.; Crovace, A.; Piccininni, A.; Cusanno, A.; Guglielmi, P.; De Napoli, L.; Serratore, G. A Structured Approach for the Design and Manufacturing of Titanium Cranial Prostheses via Sheet Metal Forming. *Metals* **2022**, *12*, 293. https://doi.org/10.3390/met12020293

Academic Editor: Mieczyslaw Jurczyk

Received: 30 December 2021
Accepted: 1 February 2022
Published: 8 February 2022

1. Introduction

The healthcare sector must continually deal with aspects of economic nature: the minimization of both the surgical timing and the complete recovery of the patient represents one of the primary objectives in this regard. This goal has led to continuous progress and innovations with the clear aim of obtaining highly performing implants for the individual patient. In light of this, one of the most important challenges involving the field of medicine is represented by the manufacture of highly customized prosthetic implants able to restore to the patient both the functionality and the natural conformation of the damaged part [1]. The need to perfectly reproduce the bone conformation of the reference patient requires a reliable procedure based on certified protocols that guarantees the perfect fitting between the prosthetic implant and the host area to be healed [2].

In this scenario, all the aspects involved are equally essential, starting from the appropriate choice of the prosthesis material, passing through the acquisition and manipulation of the damaged area geometry, up to the most appropriate manufacturing technology. With reference to the fabrication of cranial prostheses, there are several proofs of the importance deriving from the use of different Computer Aided Design (CAD) and advanced production platforms [3]. In fact, CAD tools combined with advanced patient scanning techniques have drastically transformed the process of designing prosthetic devices, effectively replacing the more traditional manual approaches based on cast and invasive techniques. In addition, these methods have made it possible to reduce costs and response times [4,5]. Moreover, typical materials adopted for biomedical applications must meet different requirements such as: (i) a certain chemical composition to avoid adverse tissue reactions, (ii) excellent resistance to degradation in the case of permanent prostheses, (iii) load-bearing capacities, (iv) capability to absorb the impacts and (v) good wear resistance for prostheses involved in reciprocal sliding [6].

In this context, metallic materials such as titanium (Ti) alloys, thanks to their capability to fully match the above-mentioned requirements, appear to be promising candidates. Such alloys, in fact, are constantly growing in terms of use in the biomedical field [7] due to their high biocompatibility, excellent resistance characteristics coupled with an appropriate elastic modulus, as well as a high resistance to corrosion (even if compared to the performing chromium-cobalt alloys) [8,9]. The best fitting between the patient's anatomy and the geometry of the prosthesis, as well as required by aesthetic aspects, is motivated by the need to minimize the risk of infection due to micromotions between the implant and the surrounding bone [10].

For all these aspects, in addition to motivating the manufacture of complex shapes (e.g., very small radii, complex profiles and, in some cases, the presence of undercuts), customized prostheses require prohibitive tolerances for most of the manufacturing processes. Innovative processes, such as Single Point Incremental Forming (SPIF) and Superplastic Forming (SPF) can be considered as viable solutions. In fact, as demonstrated in previous studies, titanium custom prostheses can be successfully manufactured by both SPF and SPIF [11,12].

During the SPF process a metal sheet is plastically deformed at a high temperature by means of a pressurized gas. Specifically, this type of manufacturing process is intended for materials capable of emphasizing their deformation properties under appropriate operating conditions both in terms of temperature and strain rate [8]. Furthermore, SPF is a cost-effective method of producing a small to medium number of complex parts obtained with expensive materials and low formability at room temperature [13]. The process can be made more competitive by adopting less expensive tools (for example ceramic dies). The process temperature and gas pressure must be suitably selected according to the geometry and the material. For this scope, a numerical/experimental approach is fundamental during the whole design process. In fact, to obtain a sound complex component, an optimized manufacturing process involves a Finite Element (FE) modelling capable to correctly predict the alloy behaviour [14–16].

SPIF is based on the idea to apply a plastic deformation on a flat sheet, clamped at its periphery, by a rotating tool which describes concentric and decreasing spirals reproducing the 3D profile of the desired geometry. In this way, the material is incrementally stretched, up to reach the final shape [17]. Additionally, in this case the process design is a critical issue since the quality of the results, in terms of minimum thickness and accuracy, are strongly affected by the tool trajectory: parameters such as tool pitch (p), wall inclination angle (α), or the 3D shape positioning with respect to the flat surface need to be properly calibrated in order to optimize the quality of the formed part [18]. For this reason, when a complex profile has to be manufactured, a numerical/experimental approach is suitable for calibrating the above-mentioned parameters. Nevertheless, SPIF is a really cheap technology for producing small batch and, mainly, single part such as in custom made prosthesis application since it does not require any additional dedicated equipment but only general-purpose tool and clamping frame [19].

The aim of the present work is to provide an effective methodology for the design and subsequent fabrication of a customized cranial prosthesis in Ti6Al4V-ELI, validated by means of in vivo tests. In particular, a numerical/experimental design method was proposed for two different innovative sheet metal forming processes (SPF and SPIF) aimed to the fabrication of the prosthetic implants.

The Computed Tomography (CT) scan images were used to generate a three-dimensional model of the sheep's damaged skull. Such geometrical acquired data were properly modified via CAD and allowed to obtain the target geometry on which the study was based. For both investigated manufacturing processes, a Finite Element approach was used to: (i) design the gas pressure (for the SPF process); (ii) define the tool trajectory (for the SPIF process). The post forming characteristics, such as thickness distributions and the shape accuracy of the prostheses, were also evaluated. Finally, the in vivo response of the prostheses manufactured using both the processes was assessed by means of histological analyses on tissues extracted after 3 and 6 months from 16 sheep in which they were implanted.

2. Materials and Methods

2.1. Investigated Materials

For all the activities involved in the present work the extra low interstitial titanium alloy Ti6Al4V-ELI (thickness: 1 mm) was used. The chemical composition of the investigated material is reported in Table 1.

Table 1. Chemical composition of the investigated Ti alloys.

Element	Al%	V%	Fe%	C%	N%	H%	O%	Ti
value	5.88	3.87	0.14	0.22	0.006	0.002	0.112	Bal.

2.2. Methodology

The whole design chain is described through the flowchart in Figure 1: starting from the CT scan of the sheep skull, the Region of Interest (ROI) was then identified, and the bone surface acquired.

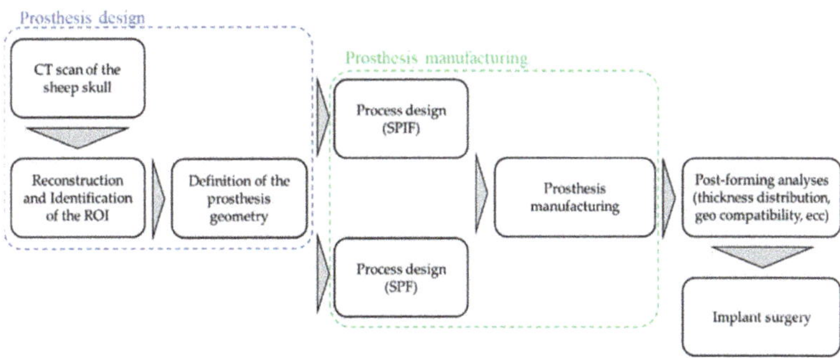

Figure 1. Schematic overview of the complete procedure for designing and producing titanium cranial prostheses.

The CAD reconstruction of the bone region (where the defect had to be intentionally created) represented the starting point for the design of the prosthesis geometry that, in turns, was the input data for the manufacturing process design by means of an FE-based approach. Prostheses were then manufactured—using both the SPF and the SPIF process—according to the results of the numerical simulations and finally implanted.

The starting point, after the CT scan, was the Digital Imaging and COmmunications in Medicine (DICOM) manipulation to obtain the 3D model of the skull. Figure 2 shows a typical overview resulting from the imported DICOM files into a dedicated software together with the reconstruction of the skull volume (in the bottom right corner).

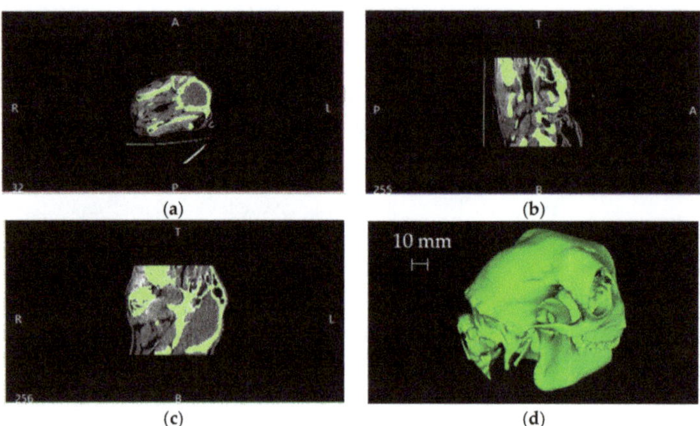

Figure 2. Schematic overview of the imported DICOM files and the resulting 3D skull volume reconstruction: (**a**) axial slice, (**b**) sagittal slice, (**c**) coronal slice and (**d**) 3D volume.

This volume was then exported in the Standard Triangulation Language (STL) format. The data, represented by a point cloud, were subsequently imported and processed with a CAD software. When required, these points were divided into significant groups and, if necessary, filtered. However, the goal was to obtain a virtual model of the portion of the skull on which the prosthesis had to be implanted (ROI), fundamental for the virtual modelling of the defect to be intentionally created. Therefore, from this point, only the surface correspondent to the ROI was considered for subsequent manipulation.

An elliptical damage geometry was defined, characterized by the dimension of the largest and smallest axes equal to 34 mm and 28 mm, respectively. The identification of the damaged zone, under the advice of the surgeon, was conducted by taking as landmarks four points: points 1 and 2 identify an area immediately adjacent to the point of attachment of the nuchal tendons to the skull, whereas points 3 and 4 identify the sagittal plane. In this way, the prosthetic geometry (also elliptical) was characterized by such a positioning that its major axis will be oriented according to the just-defined sagittal plane, and the lower point is located near the point 4, as reported in Figure 3.

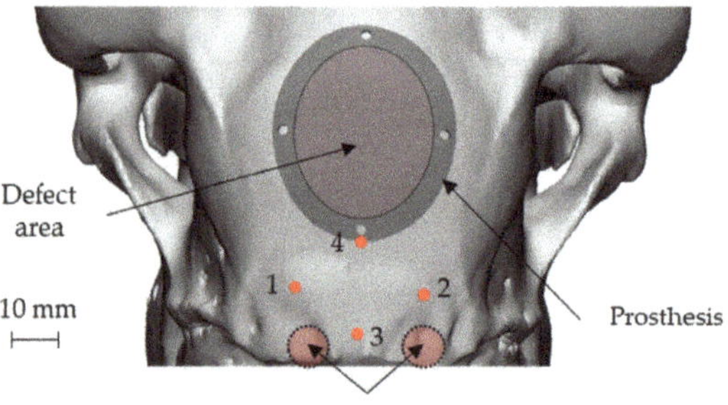

Figure 3. Position of the nuchal tendon attachment points on the skull, the 4 landmarks and representation of both the defect and the prosthetic geometry: attachment of the nuchal tendons (points 1 and 2), sagittal plane (points 3 and 4).

Subsequently, to define the CAD model of the prosthesis, the analytical surfaces close to the ROI were reconstructed starting from the obtained points. Figure 4 shows the reconstructed surface geometry compared with the original point cloud (red mesh) to verify the accuracy of the whole routine.

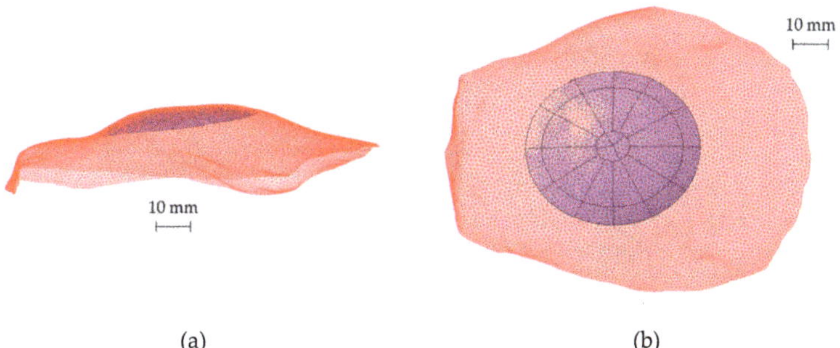

(a)   (b)

Figure 4. (**a**) Lateral view and (**b**) top view of the reconstructed analytical surface of the ROI and the original portion of the skull (red mesh) including the ROI.

The prosthesis geometry was designed as elliptical by enlarging the value of the main axes from the elliptical damage up to 42 mm and 36 mm. The final 3D model of the prosthesis was obtained by making a thickening of this surface.

The approach of skull geometry reconstruction was also used to create a customized "guiding mask", therefore perfectly respectful of bone geometry, to guide the surgeon during the intentional defect creation, and also to make the prosthesis positioning highly accurate. Since this mask was made on the specific anatomy of the skull, it minimized the error of positioning by the surgeon: in fact, being the position of the mask completely uni-vocal with respect to the cranial case, such an approach made the operations of generating the damage highly accurate and less affected by positioning errors. Figure 5 shows the guiding mask and its position on the skull together with the geometry of the defect.

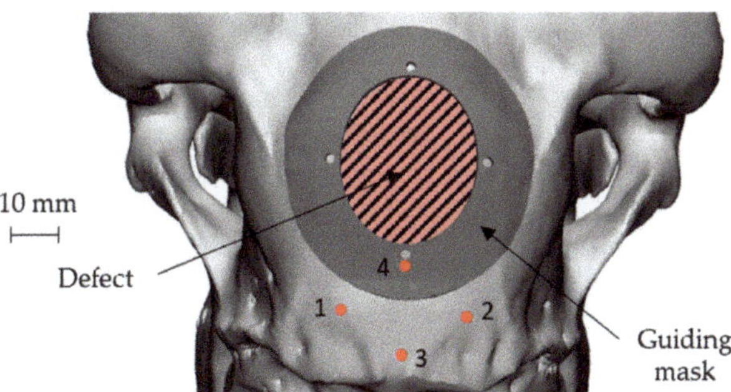

Figure 5. Geometry and positioning of the guiding mask on the skull with respect to the 4 landmarks and to the defect geometry: attachment of the nuchal tendons (points 1 and 2), sagittal plane (points 3 and 4).

Figure 6 shows the final geometry of the prosthesis placed on the skull using point 4 as reference. The four holes visible on the prosthesis coincided in position and size with those present on the guiding mask. The role of the mask, in addition to guiding the surgeon during the defect creation, could also help to create the holes for the implant anchorage, thus allowing the correct positioning of the implant (Figure 3).

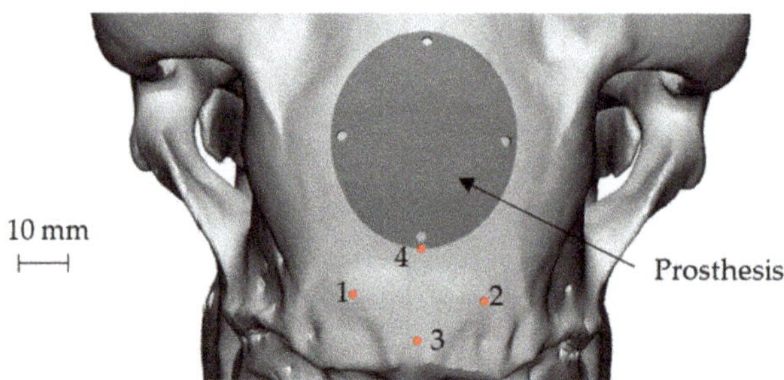

Figure 6. The final geometry of the prosthesis placed on the skull: attachment of the nuchal tendons (points 1 and 2), sagittal plane (points 3 and 4).

2.3. SPIF Process Design and Prosthesis Manufacturing

During the design of the SPIF process, attention was paid not only to the final geometry of the prosthesis but also to the best positioning of the 3D profile on the flat surface, in order to reduce the blank size and to optimize the final accuracy. The first constraint is represented by the feasibility of the process, then ensuring the forming of the part without excessive thinning while minimizing the material's waste.

The possibility to obtain the prosthesis geometry by SPIF requires the construction of a sidewall with a shape generally conical: to deform the sheet incrementally without any fracture occurrence, a vertical wall should be avoided. On the contrary, a small inclination of the sidewall would generate a lot of waste. Starting from the prosthesis surface reported in Figure 7a, the sidewall inclination and the best positioning of the prosthesis in the forming area were designed with the aid of numerical simulations. In detail, FE simulations were performed modeling the sheet as a deformable body with S4R elements and 5 integration points along the thickness, whereas the tool was modeled as a rigid body.

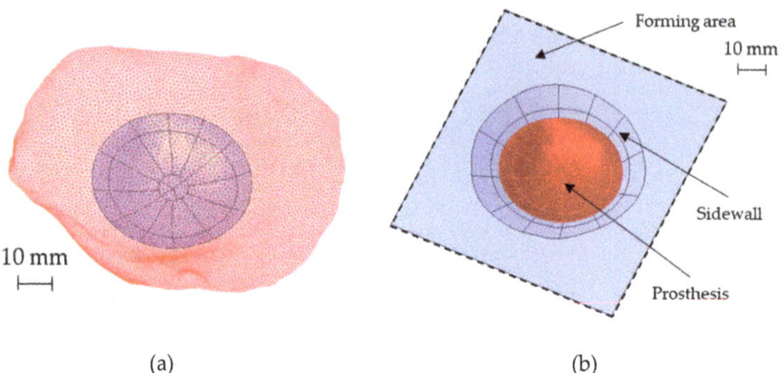

Figure 7. (**a**) The reconstructed analytical surface of the ROI and the original portion of the skull (red mesh) near the ROI, (**b**) the designed final geometry for the SPIF.

The tool path was generated with the CAD/CAM software Creo (v7, PTC, Boston, MA, USA). The periphery of the blank was pinned. The contact was modeled as surface-to-surface contact, setting a Coulomb friction coefficient equal to 0.1 between the tool and the sheet. Finally, the numerical problem was solved with the implicit integration scheme. The final geometry of the part produced by SPIF, characterized by an inclined sidewall of 30°, is reported in Figure 7b.

The above-described geometry was obtained using a hemispherical head tool (diameter: 6 mm) and setting a step depth of 0.05 mm in the CAD/ CAM software used for generating the toolpath for the CNC machine. Figure 8a shows the CAM simulation results of the SPIF geometry after the forming phase. Once completed the SPIF operation, the drilling of the four holes for the prosthesis anchoring, as well as the cutting of the outer edge, was simulated (as reported in Figure 8b).

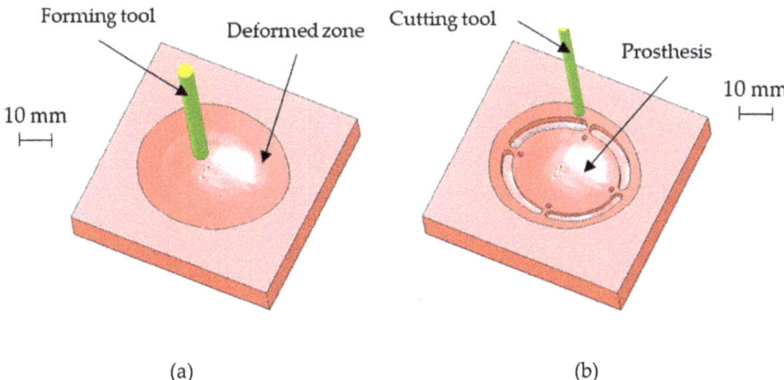

Figure 8. (**a**) CAM simulation results of the SPIF geometry after the forming phase; (**b**) final prosthesis after the CNC milling simulation.

To manufacture the prosthesis by SPIF, a Mazak Nexus 410 milling machine (Yamazaki Mazak Corporation, Oguchi, Japan) was equipped with a heating chamber able to heat the blank up to 420 °C during the forming process and to keep it constant for the whole manufacturing phase. A feed rate of 2 m/min and a spindle speed of 600 RPM were used. The final geometry of the prostheses was obtained with the same milling machine by using a milling cutter with a diameter of 4 mm and a drill bit with a diameter of 2 mm to perform the anchoring holes.

2.4. SPF Process Design and Prosthesis Manufacturing

The optimal conditions (in terms of temperature and strain rate) in which the Ti alloy exhibits the emphasized superplastic behaviour were determined in a previous experimental campaign [15]. The SPF process could be thus designed by means of an FE-based approach implementing the previously evaluated material modelling, with the aim of: (i) defining the geometry of the tool; (ii) calculating the gas pressure profile able to form the blank in the above-mentioned optimal conditions.

Starting from the CAD bone reconstruction (as discussed in Section 2.2), the SPF process was simulated using the commercial software Abaqus. The tool geometry (i.e., the die to be copied by the blank) was modelled using two different approaches, briefly labelled as convex and concave (see Figure 9).

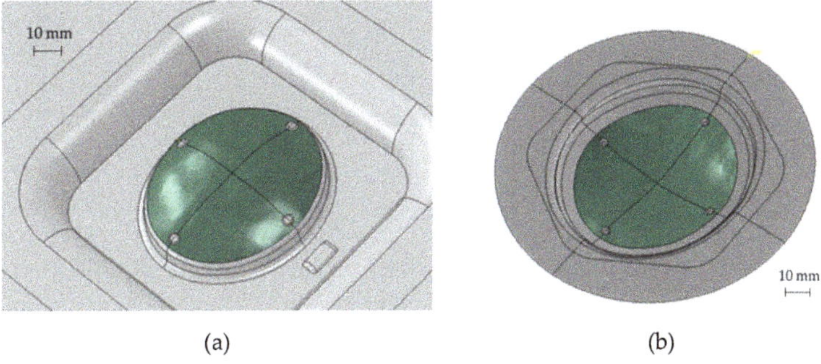

Figure 9. Different types of dies used for the SPF simulation: (**a**) convex and (**b**) concave.

As reported in Figure 9, to obtain the prosthesis geometry (green area), an additional area had to be added to deform the blank and to allow the blankholder action. In addition, 4 protrusions were included in the tool to obtain, on the formed blank, the reference positions to create the anchoring holes. Both the types of the die were modelled as a rigid body, meshing the surface with 1 mm element (R3D3 and R3D4 shell rigid element); the blank was modelled as a shell deformable body using S4R and S3R shell elements (average size equal to 0.5 mm) and considering 5 integration points in the thickness direction. The friction between the die and the blank was modelled using the Coulomb's formulation and setting the coefficient to 0.1 [20]. The gas pressure profile was calculated by means of an internal subroutine (CREEP STRAIN RATE CONTROL) by specifying that the blank portion in contact with the die surface (the one highlighted in green in Figure 9) had to be deformed under a strain rate target level of 5×10^{-4} 1/s.

As concerns the prostheses manufacturing, an INSTRON 4485 universal testing machine (INSTRON, Norwood, MA, USA) controlled by a ZwickRoell software (version, ZwickRoell, Ulm, Germany) and equipped with a specifically designed equipment was used. The experimental set-up is composed of a load cell, which allows to control the Blank Holder Force (BHF) during the tests, an upper tool which applies the BHF, and which is connected to the pneumatic circuit for the argon supply, a lower tool and the die. The tools and the die are embedded into an electric split furnace. The starting blank was a 1 mm thickness sheet with a circular shape having a radius of 75 mm. In order to simplify the extraction of the sheet after the forming process, a thin layer of graphite was applied on the external part of the sheet (the one in contact with the blankholder). The temperature of the furnace was set at 850 °C. After the target temperature was reached in the forming chamber, the blank was placed between the tool and the die, and a BHF of 22 kN was applied. Then, the argon gas was inflated into the forming chamber according to the pressure profile calculated by the FE simulations, while the BHF was kept constant by the control system for the whole forming time. As forming tool, a specific ceramic insert placed into a cylindrical steel frame, was used for each prosthesis (customized to the specific anatomy); such a ceramic was manufactured by investment casting using the pattern shown in Figure 10a (created by stereolithography) having the negative geometry of the ceramic insert.

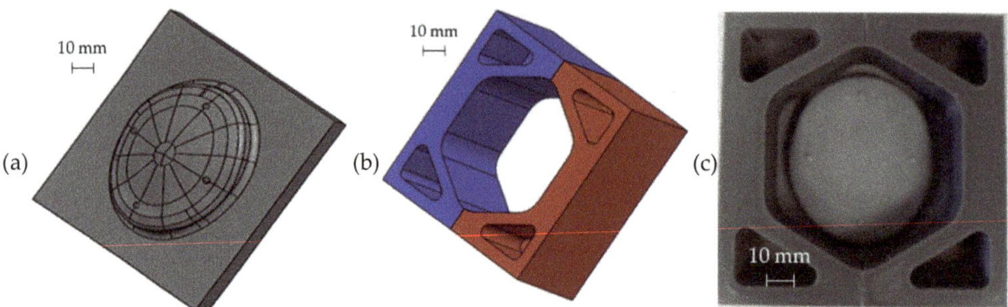

Figure 10. CAD model of (**a**) the photopolymeric base and (**b**) the upper bracket of the mould used for manufacturing the ceramic insert for the die; (**c**) assembled mould.

The mould for pouring the ceramic material (Figure 10b) was obtained by Fused Deposition and was assembled putting it on the pattern, as shown in Figure 10c. The mould is composed of two parts (as highlighted by the different colours) in order to simplify the extraction of the ceramic insert after the casting and autoclave curing operations.

2.5. Post-Forming Analyses

The post-forming analyses were aimed to evaluate the thickness distribution and the shape accuracy of the produced implants. For this purpose, a high-resolution 3D scanner was adopted and the acquisition was manipulated with the GOM Inspect Suite to reduce the noise. In order to evaluate the accuracy and the replicability of the investigated processes, 3 replications of the same prosthesis geometry were manufactured by SPF and by SPIF. Each replication was scanned with the high-resolution 3D scanner and compared with the corresponding skull geometry resulting from the DICOM image to evaluate the shape accuracy. In addition, the replications were also compared to each other in order to evaluate the variability of the process.

2.6. Implant Surgery

The in vivo study was performed at the facility of the Department of Emergencies and Organ Transplantation of the University of Bari. After the approval of the Italian Minister of Health (Approval number 654/2020 PR), 16 sheep underwent the implantation, in a blind matter, of the two kinds of prothesis. Sheep were divided in two groups: the first one that was sacrificed after 6 months from surgery and the second one after 3 months. As far as the surgery is concerned, an incision was performed at the level of the parietal bone on the skin and subcutaneous tissue (Figure 11a), the custom-made mask was positioned on the bone using screws (Figure 11b) and the ostectomy was performed via piezosurgery (Mectron, Italy) [21] (Figure 11c); the bone gusset and the mask were then removed and the former was replaced with the device that was subsequently anchored using the same holes created for the mask positioning (Figure 11d).

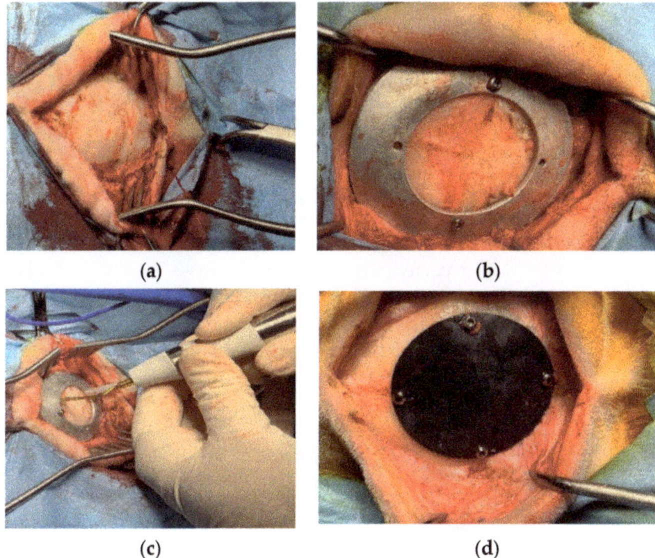

Figure 11. Surgical procedure for the in vivo studies: (**a**) exteriorizing of parietal bone; (**b**) guiding mask fixed to the bone with screws; (**c**) bone gusset removal through piezosurgery osteotomy; (**d**) device implantation and fixation with screws.

The above tissues were than sutured and the animals were awakened from anaesthesia. After the surgery, depending of the group they belonged to, animals underwent the administration of different bone markers (Tetracycline, Xilenol Orange and Calcein Blue were used for the three months group, whereas Calcein Blue, Tetracyclin Xilenol and Orange and Alizarin Red were used for the six months group). Moreover, the absence of alisteresis and screws and prosthesis mobility, as well as any periprosthetic inflammatory reaction, infections, sub cutaneal exudation and soft tissue scarring times were checked. The sheep's temperature was also regularly monitored to exclude fever.

After 24 h from the last administration, the animals were euthanized using a propofol overdose, then the heads were removed and were subjected to a CT scan to evaluate the effects of the implant on the bone. The devices were removed using the piezosurgery making a gusset of bone 1 cm from the device on all the sides, then put in formaline for seven days, washed under running water for 24 h and put in 70° ethanol to undergo histological evaluations.

3. Results and Discussion

3.1. SPIF Process Investigation and Outcomes

Since the excessive thinning of the sheet is one of the main drawbacks of the SPIF, the focus of the numerical simulation was kept on the resulting thickness distribution of the formed part. Figure 12 shows the calculated distribution of the section thickness (STH) mapped on the prosthesis at the end of the simulation.

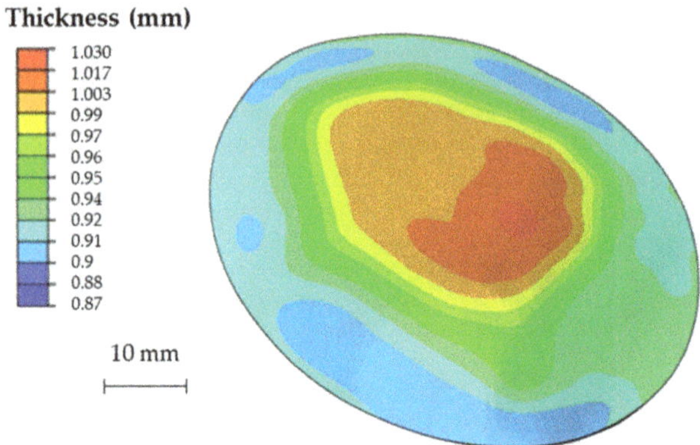

Figure 12. Numerical thickness map resulting from the FEM analysis of the SPIF process (dimensions: mm).

As expected, the minimum thickness is on the outer edge with a minimum value close to 0.9 mm, whereas the maximum thickness is located in the middle zone, being essentially equal to the initial sheet thickness (1 mm). The final geometry of the prostheses obtained by SPIF, after the cutting and drilling operations, is reported in Figure 13a; Figure 13b shows the cleaned prosthesis ready to be implanted.

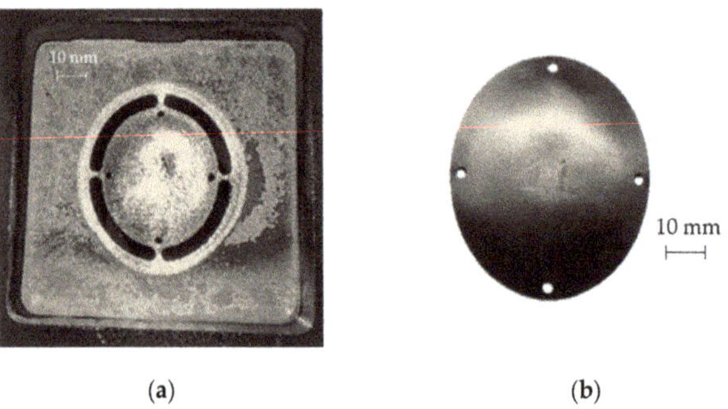

(a)   (b)

Figure 13. (a) Prosthesis after the SPIF manufacturing, cutting and drilling operations; (b) final prosthesis obtained by SPIF after cleaning operations.

Figure 14 shows the obtained thickness map on the prosthesis and the corresponding frequency of the thickness values on the whole surface. The maximum and minimum thickness values, as well as their location and distribution, are close to those predicted by the numerical simulation.

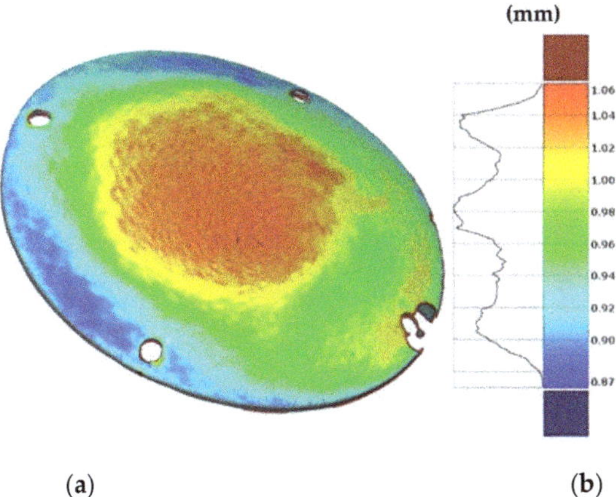

(a) (b)

Figure 14. (a) Thickness map of the prosthesis produced by SPIF and (b) frequency of the thickness values.

3.2. SPF Process Investigation and Outcomes

The FE simulation of the SPF process allowed to calculate the thickness distributions useful to check the presence of any critical region and to define the gas pressure profile to be used in the manufacturing process. In addition, it was possible to assess the key role played by the die geometry, which resulted to have a significant effect not only on the profile but also on the total forming time, as shown in Figure 15.

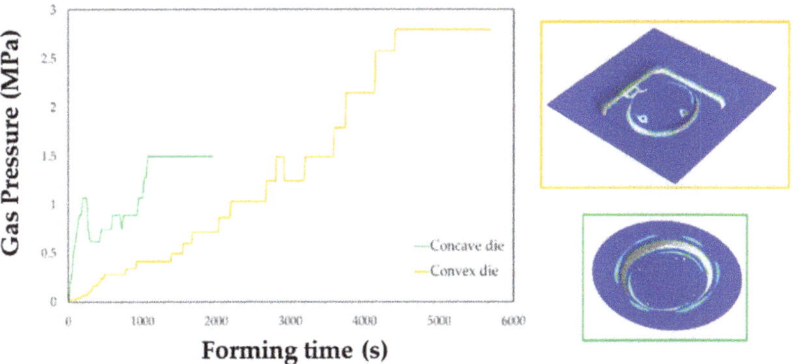

Figure 15. Gas pressure profiles obtained from the simulations of the SPF process when using the two investigated types of tools.

The maps showing the distribution of the COPEN output variable are reported for both the investigated types of tools: in fact, such an output variable describes the capability of the blank to fill the die cavity (the blue regions refer to a complete contact between the die and the blank, whereas the white ones indicate that the points are still not adjacent). A consideration that could be drawn from the presented plot is that with both the tools the part corresponding to the final prosthesis geometry was successfully copied; furthermore, by looking at the two pressure profiles, the concave geometry led to a satisfactory filling of the die cavity in a sensibly lower period of time than the other solution. Numerical results were also analysed in terms of final thickness distributions. The curves plotted in Figure 16 suggest that, within the blank portion of the final prosthesis geometry (highlighted by the grey box on each plot), both the investigated tools led to an almost uniform final distribution of thickness along the considered paths, about equal to 0.85 mm.

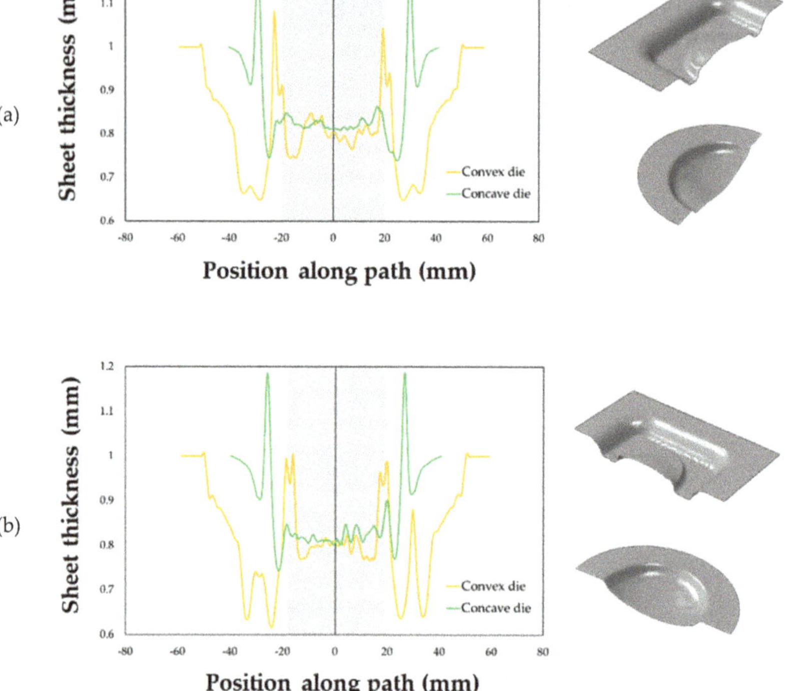

Figure 16. Thickness distributions along: (**a**) the longitudinal and (**b**) transversal symmetry paths.

Due to the large forming time reduction that it allows, the concave die was thus used for the manufacturing of all prostheses. In Figure 17 the customized ceramic insert positioned in the steel frame is shown.

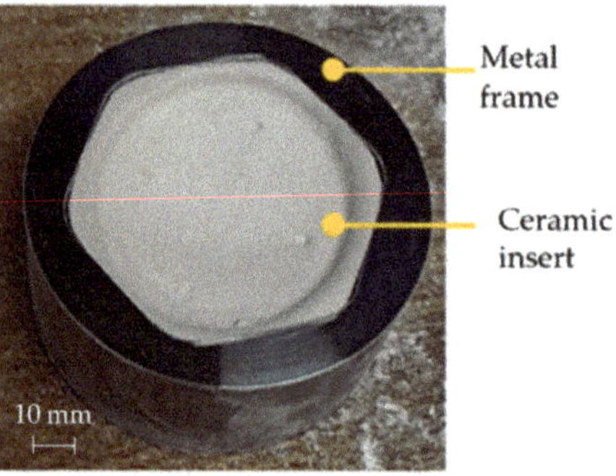

Figure 17. Die adopted during the prostheses manufacturing via SPF.

In Figure 18a the geometry of the prosthesis obtained by means of the SPF process is shown, whereas Figure 18b shows the custom prosthesis to be implanted extracted from the formed sheet by wire EDM.

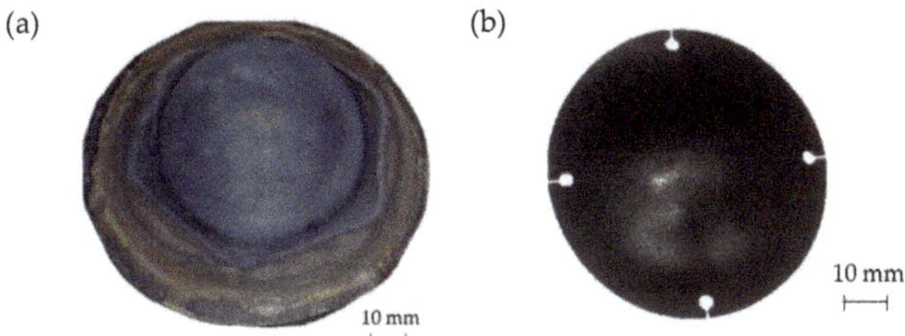

Figure 18. (**a**) Prostheses manufactured via SPF; (**b**) final geometry obtained after EDM cutting.

Finally, Figure 19 shows the thickness distribution mapped on the prosthesis obtained via SPF and the corresponding frequency of the thickness values on the whole surface. It is worthy of notice that the prosthesis was characterized by a homogeneous thickness distribution, ranging between 0.91 mm and 0.95 mm.

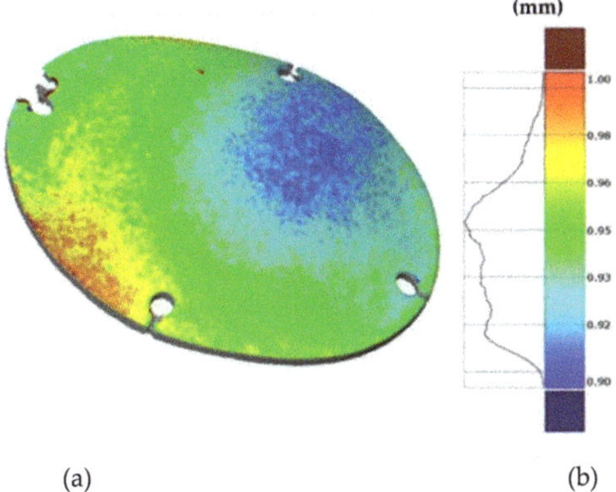

Figure 19. (**a**) Thickness map of the prosthesis produced by SPF; (**b**) frequency of thickness values.

3.3. Shape Accuracy and Replicability

The three replications of the same prosthesis manufactured by both the investigated processes were analysed one to each other and with respect to the initial CAD surface of the skull. Tables 2 and 3 show the comparison of the scanned data concerning the three replications obtained by SPIF and SPF in terms of average alignment error.

It can be seen that the investigated forming processes are characterized by a low variability, since in both the cases the higher value of the average error between the replications is below 0.01 mm.

At the same time, the SPF process reveals to be able to produce prostheses characterised by higher accuracy (the average error between the deformed geometry and the CAD model is close to zero). However, it should be noted that such a result is determined by the adoption of a female die which, on the contrary, was not necessary for the SPIF process.

It should be also noted that, even without a die, the SPIF process allowed us to obtain a prosthesis characterised by an average error between the deformed geometry and the CAD model very small (about 0.12 mm). The proposed results confirm the effectiveness of the investigated manufacturing processes in ensuring a high dimensional accuracy of the formed blanks: SPIF prostheses' deviations are comparable to those achieved after the manufacturing of a TA1 cranial implant [22]. Both the investigated processes were accurate, ensuring lower deviations if compared to the accuracy achievable using the same alloy but more conventional [23] or, alternatively, an additive manufacturing approach [24]. The level of correspondence between the formed and the designed geometry is comparable, in the case of the SPIF, or even lower, in the case of the SPF, to the one reported in the literature regarding the manufacturing of PEEK [25] and PMMA [26] customized implants by 3D printing.

Table 2. Average error of the three replications obtained by SPIF.

Compared Geometries	Average Error (mm)	Map
SPIF #1 vs. SPIF #2	0.0150	
SPIF #1 vs. SPIF #3	0.0193	
SPIF #2 vs. SPIF #3	0.0160	
SPIF #2 vs. CAD	0.1202	

Table 3. Average error of the three SPF replications.

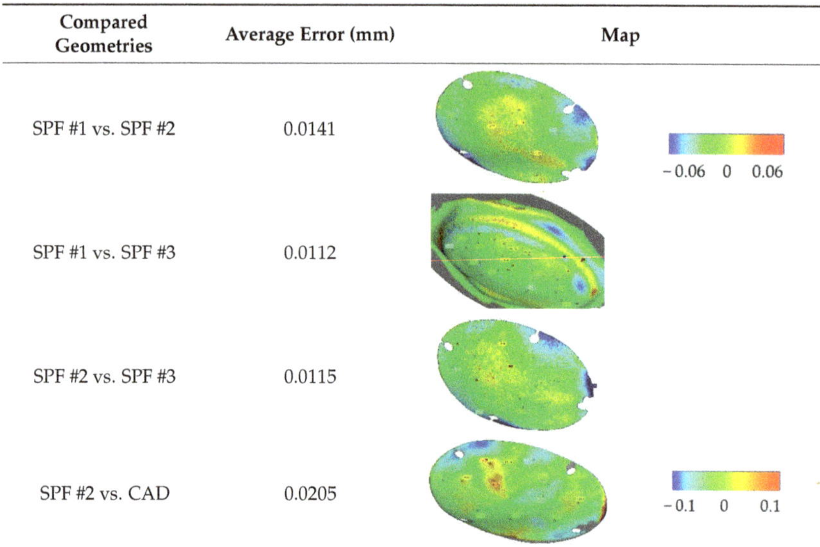

Compared Geometries	Average Error (mm)	Map
SPF #1 vs. SPF #2	0.0141	
SPF #1 vs. SPF #3	0.0112	
SPF #2 vs. SPF #3	0.0115	
SPF #2 vs. CAD	0.0205	

3.4. In Vivo Behaviour of the Prostheses

The adopted procedure for the design of both the prosthesis and the surgery tools (the guiding mask) revealed to be effective since, after the surgery, in all the treated animals there were no side effects noticed in the prothesis area, the reparation of the soft tissue was normal and according to the scarring times; moreover, the body temperature was normal in the post-operative period and all the animals resumed the normal behaviour immediately.

The border of the region that was obtained by the ostectomy was then analysed evaluating the presence of thermal damages on the bone and the presence of damages on the other nervous tissue and the dura madre around the ostectomy area.

Both the types of the implant produced using the two investigated processes resulted to fit suitably the anatomy of the specific animal, thus avoiding any modelling of the prosthesis in the surgery room.

In addition, the implanted prostheses revealed to be much thinner than the ones which could be produced by both standard (machining) and innovative (Additive Manufacturing) production technologies, being at the same time enough resistant.

After 3 and 6 months from the surgery, the sheep were sacrificed and the prostheses extracted in order to be analysed. Figure 20 shows one of the prostheses (from one of the 3 months sacrificed sheep) and the bone gusset removed post mortem.

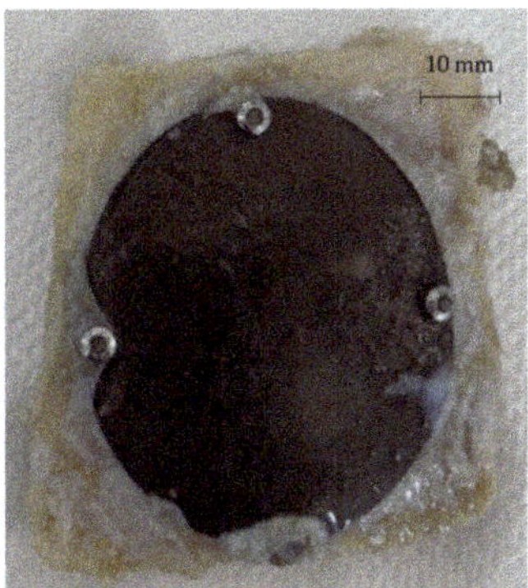

Figure 20. Prosthesis and bone gusset removed post mortem.

4. Conclusions

In this study an original methodology based on a coupled mechanical and surgical approach has been proposed. The procedure combines the well assessed prosthesis reconstruction procedures, based on the conversion of DICOM imagine into CAD model, thus ensuring the proper definition of the implant geometry. The manufacturing of an innovative guiding mask can be regarded as an innovative step to ensure the correct and replicable positioning of the implant during the surgery, thus minimizing any mismatch with the per-implant bone. The respect of the bone geometry was then ensured by relying on two innovative and flexible manufacturing processes: the scans on the formed blanks demonstrated not only a high level of the geometry replicability (few hundredths of mm) but also very low deviations from the designed CAD geometry, even lower than what is achievable with alternative manufacturing approaches.

The effectiveness of the design and manufacturing chain, applied to the case study of a customized prosthesis for sheep, was finally evaluated carrying out the in vivo tests. The absence of any side effect (as evidence of the high mechanical fixation of the prosthesis on the surrounding bones) demonstrates that the full application of this methodology ensures both high accuracy and significant replicability of the results, validating the effectiveness of the whole approach.

Author Contributions: Formal analysis: A.C. (Alberto Crovace), A.P., A.C. (Angela Cusanno), P.G., L.D.N. and G.S.; Funding acquisition: G.P. and G.A.; Investigation: A.P., A.C. (Angela Cusanno), P.G., A.C. (Alberto Crovace) and G.S.; Methodology: G.P. and G.A.; Supervision: G.P.; Writing—original draft: A.P., A.C. (Angela Cusanno), P.G. and G.S.; Writing—review & editing: G.P. and G.A. All authors have read and agreed to the published version of the manuscript.

Funding: This research was funded by the Italian Ministry of University and Research through the call Proof of Concept 2018 (acronym: FORMAE-BIO), grant number POC01_00097.

Institutional Review Board Statement: The study was conducted according to the guidelines of the Declaration of Helsinki, and approved by the Ethics Committee (protocol code 654/2020 PR).

Informed Consent Statement: Not applicable.

Data Availability Statement: Not applicable.

Acknowledgments: The authors wish to thank Tomaso Villa for his contribution in acquiring the funds.

Conflicts of Interest: The authors declare no conflict of interest; the funders had no role in the design of the study; in the collection, analyses, or interpretation of data; in the writing of the manuscript, or in the decision to publish the results.

References

1. Castelan, J.; Schaeffer, L.; Daleffe, A.; Fritzen, D.; Salvaro, V.; Da Silva, F.P. Manufacture of custom-made cranial implants from DICOM® images using 3D printing, CAD/CAM technology and incremental sheet forming. *Rev. Bras. Eng. Biomed.* **2014**, *30*, 265–273. [CrossRef]
2. Mallya, P.K.; Juneja, M. Rapid prototyping of orthopedic implant materials for cranio-facial reconstruction: A survey. *Mater. Today Proc.* **2021**, *45*, 5207–5213. [CrossRef]
3. Saldarriaga Design and Manufacturing of a Custom Skull Implant. *Am. J. Eng. Appl. Sci.* **2011**, *4*, 169–174. [CrossRef]
4. Mohammed, M.I.; Tatineni, J.; Cadd, B.; Peart, G.; Gibson, I. Advanced auricular prosthesis development by 3D modelling and multi-material printing. *KnE Eng.* **2017**, *2*, 37. [CrossRef]
5. Tanveer, W.; Ridwan-Pramana, A.; Molinero-Mourelle, P.; Koolstra, J.H.; Forouzanfar, T. Systematic review of clinical applications of cad/cam technology for craniofacial implants placement and manufacturing of nasal prostheses. *Int. J. Environ. Res. Public Health* **2021**, *18*, 3756. [CrossRef]
6. Kiradzhiyska, D.D.; Mantcheva, R.D. Overview of Biocompatible Materials and Their Use in Medicine. *Folia Med. (Plovdiv).* **2019**, *61*, 34–40. [CrossRef]
7. Cai, D.; Zhao, X.; Yang, L.; Wang, R.; Qin, G.; Chen, D.F.; Zhang, E. A novel biomedical titanium alloy with high antibacterial property and low elastic modulus. *J. Mater. Sci. Technol.* **2021**, *81*, 13–25. [CrossRef]
8. Jackson, M. *Superplastic Forming of Advanced Metallic Materials*; Elsevier: Amsterdam, The Netherlands, 2011; ISBN 9781845697532.
9. Disegi, J.A. Titanium alloys for fracture fixation implants. *Injury* **2000**, *31* (Suppl. S4), 14–17. [CrossRef]
10. Wazen, R.M.; Currey, J.A.; Guo, H.; Brunski, J.B.; Helms, J.A.; Nanci, A. Micromotion-induced strain fields influence early stages of repair at bone-implant interfaces. *Acta Biomater.* **2013**, *9*, 6663–6674. [CrossRef]
11. Piccininni, A.; Gagliardi, F.; Guglielmi, P.; De Napoli, L.; Ambrogio, G.; Sorgente, D.; Palumbo, G. Biomedical Titanium alloy prostheses manufacturing by means of Superplastic and Incremental Forming processes. In *MATEC Web of Conferences*; EDP Sciences: Les Ulis, France, 2016; Volume 80.
12. Cheng, Z.; Li, Y.; Xu, C.; Liu, Y.; Ghafoor, S.; Li, F. Incremental sheet forming towards biomedical implants: A review. *J. Mater. Res. Technol.* **2020**, *9*, 7225–7251. [CrossRef]
13. Limited, W.P. Processing and machining of aerospace metals. *Introd. Aerosp. Mater.* **2012**, 154–172. [CrossRef]
14. Mosleh, A.O.; Mikhaylovskaya, A.V.; Kotov, A.D.; Kwame, J.S. Experimental, modelling and simulation of an approach for optimizing the superplastic forming of Ti-6%Al-4%V titanium alloy. *J. Manuf. Process.* **2019**, *45*, 262–272. [CrossRef]
15. Sorgente, D.; Palumbo, G.; Piccininni, A.; Guglielmi, P.; Aksenov, S.A. Investigation on the thickness distribution of highly customized titanium biomedical implants manufactured by superplastic forming. *CIRP J. Manuf. Sci. Technol.* **2018**, *20*, 29–35. [CrossRef]
16. Guglielmi, P.; Cusanno, A.; Bagudanch, I.; Centeno, G.; Ferrer, I.; Garcia-Romeu, M.L.; Palumbo, G. Experimental and numerical analysis of innovative processes for producing a resorbable cheekbone prosthesis. *J. Manuf. Process.* **2021**, *70*, 1–14. [CrossRef]
17. Ambrogio, G.; Gagliardi, F.; Chamanfar, A.; Misiolek, W.Z.; Filice, L. Induction heating and cryogenic cooling in single point incremental forming of Ti-6Al-4V: Process setup and evolution of microstructure and mechanical properties. *Int. J. Adv. Manuf. Technol.* **2017**, *91*, 803–812. [CrossRef]
18. Filice, L.; Ambrogio, G.; Gaudioso, M. Optimised tool-path design to reduce thinning in incremental sheet forming process. *Int. J. Mater. Form.* **2013**, *6*, 173–178. [CrossRef]
19. Ambrogio, G.; Palumbo, G.; Sgambitterra, E.; Guglielmi, P.; Piccininni, A.; De Napoli, L.; Villa, T.; Fragomeni, G. Experimental investigation of the mechanical performances of titanium cranial prostheses manufactured by super plastic forming and single-point incremental forming. *Int. J. Adv. Manuf. Technol.* **2018**, *98*, 1489–1503. [CrossRef]

20. Sorgente, D.; Palumbo, G.; Piccininni, A.; Guglielmi, P.; Tricarico, L. Modelling the superplastic behaviour of the Ti6Al4V-ELI by means of a numerical/experimental approach. *Int. J. Adv. Manuf. Technol.* **2017**, *90*, 1–10. [CrossRef]
21. Crovace, A.M.; Luzzi, S.; Lacitignola, L.; Fatone, G.; Lucifero, A.G.; Vercellotti, T.; Crovace, A. Minimal invasive piezoelectric osteotomy in neurosurgery: Technic, applications, and clinical outcomes of a retrospective case series. *Vet. Sci.* **2020**, *7*, 68. [CrossRef]
22. Lu, B.; Ou, H.; Shi, S.Q.; Long, H.; Chen, J. Titanium based cranial reconstruction using incremental sheet forming. *Int. J. Mater. Form.* **2016**, *9*, 361–370. [CrossRef]
23. Peel, S.; Eggbeer, D.; Burton, H.; Hanson, H.; Evans, P.L. Additively manufactured versus conventionally pressed cranioplasty implants: An accuracy comparison. *Proc. Inst. Mech. Eng. Part H J. Eng. Med.* **2018**, *232*, 949–961. [CrossRef] [PubMed]
24. Moiduddin, K.; Mian, S.H.; Umer, U.; Alkhalefah, H. Fabrication and analysis of a Ti6Al4V implant for cranial restoration. *Appl. Sci.* **2019**, *9*, 2513. [CrossRef]
25. Berretta, S.; Evans, K.; Ghita, O. Additive manufacture of PEEK cranial implants: Manufacturing considerations versus accuracy and mechanical performance. *Mater. Des.* **2018**, *139*, 141–152. [CrossRef]
26. Chamo, D.; Msallem, B.; Sharma, N.; Aghlmandi, S.; Kunz, C.; Thieringer, F.M. Accuracy assessment of molded, patient-specific polymethylmethacrylate craniofacial implants compared to their 3D printed originals. *J. Clin. Med.* **2020**, *9*, 832. [CrossRef] [PubMed]

 *metals*

Article

Analysis of Strain-Hardening Viscoplastic Wide Sheets Subject to Bending under Tension

Sergei Alexandrov [1,2,*] and Elena Lyamina [2]

[1] Metal Forming Department, Samara National Research University, 34 Moskovskoye Shosse, 334086 Samara, Russia
[2] Laboratory for Technological Processes, Ishlinsky Institute for Problems in Mechanics RAS, 101-1 Prospect Vernadskogo, 119526 Moscow, Russia; lyamina@inbox.ru
* Correspondence: sergei_alexandrov@spartak.ru

Abstract: The present paper provides an accurate solution for finite plane strain bending under tension of a rigid/plastic sheet using a general material model of a strain-hardening viscoplastic material. In particular, no restriction is imposed on the dependence of the yield stress on the equivalent strain and the equivalent strain rate. A special numerical procedure is necessary to solve a non-standard ordinary differential equation resulting from the analytic treatment of the boundary value problem. A numerical example illustrates the general solution assuming that the tensile force vanishes. This numerical solution demonstrates a significant effect of the parameter that controls the loading speed on the bending moment and the through-thickness distribution of stresses.

Keywords: strain-hardening; viscoplasticity; bending; semi-analytic solution

Citation: Alexandrov, S.; Lyamina, E. Analysis of Strain-Hardening Viscoplastic Wide Sheets Subject to Bending under Tension. *Metals* **2022**, *12*, 118. https://doi.org/10.3390/met12010118

Academic Editor: Gabriel Centeno Báez

Received: 17 December 2021
Accepted: 4 January 2022
Published: 7 January 2022

Publisher's Note: MDPI stays neutral with regard to jurisdictional claims in published maps and institutional affiliations.

1. Introduction

Bending is frequently used as a test for determining various material properties. A four-point bend apparatus was used in [1] for getting stress–strain curves in the tension and compression of three materials: beryllium, cast iron, and copper. No significant difference between these curves and the corresponding curves from the conventional uniaxial tension and compression tests was found. Papers [2,3] applied the same method for determining the stress–strain curves for annealed copper and cement-based composites, respectively. The theoretical solution proposed in [1] was modified in [4]. Then, the method was used for determining the stress–strain curves for pure magnesium and S45C steel. Paper [5] studied the elastic/plastic behavior and the failure of CLARE laminates in bending experimentally. An elastic/plastic material model was proposed using these experimental data. Lightweight-aggregate concrete beams were tested in bending to failure in [6]. This work emphasized the location of the neutral axis at failure. Adhesively bonded bending specimens were employed in [7] for determining the bilinear elastic/plastic shear stress–strain behavior of acrylic adhesives. Hybrid sandwich structures were tested in bending in [8] to construct failure mode maps.

Experimental data should be usually supplemented with theoretical solutions for identifying material models. Most semi-analytic solutions have dealt with strain hardening materials [9–12]. Paper [12] also accounted for plastic anisotropy and tension/compression asymmetry. The representation of strain or work hardening in the strain reversal region has been simplified in these works. Paper [13] developed a semi-analytic method to treat this region without any simplification. This method was combined with various constitutive equations to describe pure bending and bending under tension of wide sheets [14]. Under certain conditions, viscoplastic or strain-hardening viscoplastic constitutive equations are required to analyze bending processes adequately [15–17]. Many material models of this type are available in the literature [18–24]. For identifying constitutive equations, it is desirable to have a theoretical solution for a general material model. The motivation of the present paper is to provide a semi-analytic solution of the same level of complexity as available solutions but for quite a general material model. It is believed that such a solution is useful for identifying constitutive equations. The solution is based on the general method proposed in [13].

The solution found is facilitated by using the equivalent strain instead of the space coordinate as an independent variable. Similar changes of independent variables have been successfully used in several works [25–27]. In particular, the process of bending was analyzed in [26]. However, in all these cases, it has been assumed that the yield stress depends on the single kinematic quantity, the equivalent plastic strain. The novelty of the present solution is that the yield stress is an arbitrary function of the equivalent strain and the equivalent strain rate. An important property of the new solution is that process parameters depend on the loading speed. In addition, the solution shows that some assumptions used in simplified formulations are not valid and corresponding numerical results may lead to the misinterpretation of experimental data.

2. Statement of the Problem

What is considered is the process of bending under tension due to large strains under plane strain conditions in the classical formulation proposed in [28], where a model of rigid perfectly plastic material was adopted. A schematic diagram of the process is shown in Figure 1. The cross section of a sheet is rectangular at the initial instant. The initial thickness of the sheet is H. As the deformation proceeds, the initially straight lines A_1B_1 and C_1D_1 become circular arcs AB and CD, respectively. The initially straight lines B_1C_1 and A_1D_1 remain straight (lines BC and AD, respectively). The radii of circular arcs AB and CD are denoted as r_{AB} and r_{CD}, respectively. A Cartesian coordinate system (x, y) is introduced. Its x-axis coincides with the axis of symmetry of the process. Its origin coincides with the intersection of the outer radius of the sheet and the axis of symmetry. Due to symmetry, it is sufficient to consider the domain $y \geq 0$. The orientation of line BC relative to the x-axis is θ_0. The actual distribution of stresses over BC is replaced with the bending moment M and the tensile force F. Both are per unit length. The surface AB is traction-free. Equilibrium requires that the tensile force is compensated with pressure over the surface CD. It is assumed that this pressure is uniformly distributed.

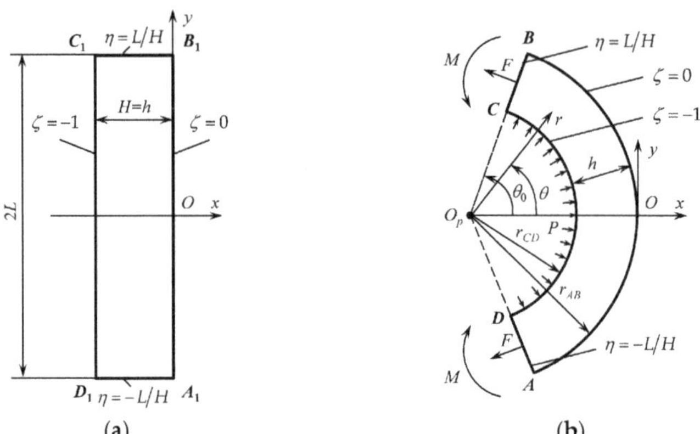

Figure 1. Schematic diagram of the bending process: (**a**) initial configuration and (**b**) intermediate and final configurations.

Then,

$$P = \frac{F}{r_{CD}}.$$

(1)

The constitutive equations include an isotropic pressure-independent yield criterion (i.e., the linear invariant of the stress tensor does not affect plastic yielding) and its associated flow rule. The elastic portion of the strain tensor is neglected. The present paper assumes that the tensile yield stress is a function of the equivalent strain, ε_{eq}, and the equivalent strain rate, ξ_{eq}, obtained by fitting any experimental data. Then, the plane-strain yield criterion can be represented as

$$|\sigma_1 - \sigma_2| = \frac{2\sigma_0}{\sqrt{3}}\Phi\left(\varepsilon_{eq},\ \xi_{eq}\right). \tag{2}$$

Here σ_1 and σ_2 are the principal stresses, σ_0 is a reference stress, and $\Phi\left(\varepsilon_{eq},\ \xi_{eq}\right)$ is an arbitrary function of its arguments. The associated flow rule is equivalent to the equation of incompressibility and the condition that the stress and strain rate tensors are coaxial. In the case under consideration, the equivalent strain rate is determined as

$$\xi_{eq} = \sqrt{\frac{2}{3}}\sqrt{\xi_1^2 + \xi_2^2} \tag{3}$$

where ξ_1 and ξ_2 are the principal strain rates. The equivalent plastic strain is given by

$$\frac{d\varepsilon_{eq}}{dt} = \xi_{eq}. \tag{4}$$

here d/dt denotes the convected derivative.

3. Kinematics

The approach for solving plane-strain bending problems proposed in [13] consists of two major steps. The first step describes the kinematics of the process. This step is the same for all isotropic incompressible materials. The second step concerns the determination of the stresses, bending moment, and tensile force. This step depends on the constitutive equations chosen. The present section outlines the basic relations from the first step derived in [13]. These relations are necessary for the second step.

Let $(\zeta,\ \eta)$ be the Lagrangian coordinate system satisfying the conditions $x = H\zeta$ and $y = H\eta$ at the initial instant. Then, $\zeta = 0$ on AB and $\zeta = -1$ on CD throughout the process of deformation (Figure 1). The transformation equations between the $(\zeta,\ \eta)-$ and $(x,\ y)-$ coordinate systems are

$$\frac{x}{H} = \sqrt{\frac{\zeta}{a} + \frac{s}{a^2}}\cos(2a\eta) - \frac{\sqrt{s}}{a} \quad \text{and} \quad \frac{y}{H} = \sqrt{\frac{\zeta}{a} + \frac{s}{a^2}}\sin(2a\eta). \tag{5}$$

Here a is a time-like variable and s is a function of a. At the initial instant, $a = 0$. The function $s(a)$ should be found from the solution. This function must satisfy the condition

$$s = \frac{1}{4} \tag{6}$$

at $a = 0$. The coordinate lines of the Lagrangian coordinate system coincide with principal strain rate trajectories [13]. Then, the coordinate lines of the Lagrangian coordinate system coincide with principal stress trajectories for the coaxial models. Therefore, the contour of the cross-section is free of shear stresses throughout the process of deformation.

It is convenient to introduce a cylindrical coordinate system $(r,\ \theta)$ by the transformation equations:

$$\frac{x}{H} + \frac{\sqrt{s}}{a} = \frac{r\cos\theta}{H} \quad \text{and} \quad \frac{y}{H} = \frac{r\sin\theta}{H}. \tag{7}$$

Equations (5) and (7) combine to give

$$\frac{r}{H} = \sqrt{\frac{\zeta}{a} + \frac{s}{a^2}} \quad \text{and} \quad \theta = 2a\eta. \tag{8}$$

Therefore, the radial, σ_r, and circumferential, σ_θ, stresses are the principal stresses. In what follows, it is assumed that $\sigma_r \equiv \sigma_1$ and $\sigma_\theta \equiv \sigma_2$. Accordingly, $\xi_r \equiv \xi_1$ and $\xi_\theta \equiv \xi_2$. Here ξ_r is the radial strain rate and ξ_θ is the circumferential strain rate. It follows from (8) that

$$\frac{r_{AB}}{H} = \frac{\sqrt{s}}{a} \quad \text{and} \quad \frac{r_{CD}}{H} = \sqrt{\frac{s}{a^2} - \frac{1}{a}}. \tag{9}$$

It is straightforward to find the principal strain rates using (5). In particular,

$$\xi_r = -\frac{(\zeta + ds/da)}{2(\zeta a + s)}\frac{da}{dt} \quad \text{and} \quad \xi_\theta = \frac{(\zeta + ds/da)}{2(\zeta a + s)}\frac{da}{dt}, \tag{10}$$

Equations (3) and (10) combine to give

$$\xi_{eq} = \frac{|\zeta + ds/da|}{\sqrt{3}(\zeta a + s)}\frac{da}{dt}. \tag{11}$$

The equivalent strain rate vanishes at the neutral line. Therefore, it follows from (11) that the $\zeta-$ coordinate of the neutral line is

$$\zeta = \zeta_n = -\frac{ds}{da}. \tag{12}$$

The neutral line moves through the material as the deformation proceeds. It is known that it moves towards CD in the case of strain-hardening materials. Assume that it is true in the case of the constitutive equations under consideration. Specific calculations below will verify this assumption. The initial condition for Equation (4) is

$$\varepsilon_{eq} = 0 \tag{13}$$

at $a = 0$. Let ζ_{n0} be the value of ζ_n at $a = 0$. It is necessary to consider three regions for determining the equivalent strain. These regions are: $0 \geq \zeta \geq \zeta_{n0}$ (Region 1), $\zeta_{n0} \geq \zeta \geq \zeta_n$ (Region 2), and $\zeta_n \geq \zeta \geq 0$ (Region 3). In Region 1, $\zeta - \zeta_n = \zeta + ds/da \geq 0$ throughout the process of deformation. Therefore, one can substitute (11) into (4) and immediately integrate the resulting equation using (13) and (6) to get

$$\varepsilon_{eq} = \frac{1}{\sqrt{3}}\ln[4(\zeta a + s)]. \tag{14}$$

In Region 3, $\zeta - \zeta_n = \zeta + ds/da \leq 0$ throughout the process of deformation. Therefore, one can substitute (11) into (4) and immediately integrate the resulting equation using (13) and (6) to get

$$\varepsilon_{eq} = -\frac{1}{\sqrt{3}}\ln[4(\zeta a + s)]. \tag{15}$$

In Region 2, $\zeta - \zeta_n = \zeta + ds/da \geq 0$ instantly. However, strain reversal occurs in this region. Consider any $\zeta-$ curve of Region 2. Let a_c be the value of a at which this curve coincides with the neutral line and s_c is the corresponding value of s. Evidently, both a_c and s_c are functions of ζ. Then, the equivalent strain in Region 2 is given by

$$\varepsilon_{eq} = \frac{1}{\sqrt{3}}\ln\left\{\frac{\zeta a + s}{4[\zeta a_c(\zeta) + s_c(\zeta)]^2}\right\}. \tag{16}$$

4. Stress Solution

The only stress equilibrium equation that is not satisfied automatically in the cylindrical coordinate system is

$$\frac{\partial \sigma_r}{\partial r} + \frac{\sigma_r - \sigma_\theta}{r} = 0. \tag{17}$$

The yield criterion (2) can be represented as

$$\sigma_\theta - \sigma_r = \frac{2\beta\sigma_0}{\sqrt{3}}\Phi\left(\varepsilon_{eq},\,\xi_{eq}\right). \tag{18}$$

Here and in what follows, $\beta = +1$ in regions 1 and 2, and $\beta = -1$ in region 3. Equations (17) and (18) combine to give

$$\frac{\partial\sigma_r}{\sigma_0\partial r} = \frac{2\beta}{\sqrt{3}r}\Phi\left(\varepsilon_{eq},\,\xi_{eq}\right). \tag{19}$$

Using (8), one can replace differentiation with respect to r with differentiation with respect to ζ. As a result,

$$\frac{\partial\sigma_r}{\sigma_0\partial\zeta} = \frac{\beta a}{\sqrt{3}(\zeta a + s)}\Phi\left(\varepsilon_{eq},\,\xi_{eq}\right). \tag{20}$$

One can rewrite Equation (11) as

$$\chi = \frac{\beta\gamma(\zeta + ds/da)}{\sqrt{3}(\zeta a + s)}. \tag{21}$$

Here $\chi = \xi_{eq}/\xi_0$, ξ_0 is a characteristic strain rate, and $\gamma = \xi_0^{-1}da/dt$. Equations (14) and (15) can be rewritten as

$$\varepsilon_{eq} = \frac{\beta}{\sqrt{3}}\ln[4(\zeta a + s)]. \tag{22}$$

One can eliminate ζ in (21) using (22). As a result,

$$\chi = \frac{\beta\gamma\left[\exp\left(\sqrt{3}\beta\varepsilon_{eq}\right) - 4s + 4ads/da\right]}{\sqrt{3}a\exp\left(\sqrt{3}\beta\varepsilon_{eq}\right)}. \tag{23}$$

Using (22), one can replace differentiation with respect to ζ with differentiation with respect to ε_{eq}. Then, Equation (20) becomes

$$\frac{\partial\sigma_r}{\sigma_0\partial\varepsilon_{eq}} = \Phi\left(\varepsilon_{eq},\,\xi_{eq}\right). \tag{24}$$

Using (23), one can represent the function $\Phi\left(\varepsilon_{eq},\,\xi_{eq}\right)$ as

$$\Phi\left(\varepsilon_{eq},\,\xi_{eq}\right) = \Phi\left[\varepsilon_{eq},\,\xi_0\chi\left(\varepsilon_{eq}\right)\right] = \Lambda\left(\varepsilon_{eq}\right). \tag{25}$$

Equations (24) and (25) combine to give

$$\frac{\partial\sigma_r}{\sigma_0\partial\varepsilon_{eq}} = \Lambda\left(\varepsilon_{eq}\right). \tag{26}$$

Let ε_1 be the value of ε_{eq} on surface AB (Figure 1). It follows from (14) that

$$\varepsilon_1 = \frac{1}{\sqrt{3}}\ln(4s). \tag{27}$$

Since $\sigma_r = 0$ on surface AB, the distribution of the radial stress in region 1 is determined from (26) and (27) as

$$\frac{\sigma_r}{\sigma_0} = \int_{\varepsilon_1}^{\varepsilon_{eq}} \Lambda_1(\mu)d\mu. \tag{28}$$

Here μ is a dummy variable of integration and $\Lambda_1(\varepsilon_{eq})$ denotes the function $\Lambda(\varepsilon_{eq})$ when $\beta = +1$ in (23). The value of ε_{eq} at $\zeta = \zeta_{n0}$ is determined from (14) as

$$\varepsilon_0 = \frac{1}{\sqrt{3}}\ln[4(\zeta_{n0}a + s)].\tag{29}$$

Then, the radial stress at $\zeta = \zeta_{n0}$ is found from (28) in the form

$$\frac{\sigma_r^{(0)}}{\sigma_0} = \int_{\varepsilon_1}^{\varepsilon_0}\Lambda_1(\varepsilon_{eq})d\varepsilon_{eq}.\tag{30}$$

Let ε_3 be the value of ε_{eq} on surface CD (Figure 1). It follows from (15) that

$$\varepsilon_3 = -\frac{1}{\sqrt{3}}\ln[4(s - a)].\tag{31}$$

Using (9), one can represent (1) as

$$p = \frac{P}{\sigma_0} = \frac{fa}{\sqrt{s - a}}\tag{32}$$

where $f = F/(\sigma_0 H)$. It follows from (26), (31), and (32) that the distribution of the radial stress in region 3 is

$$\frac{\sigma_r}{\sigma_0} = \int_{\varepsilon_3}^{\varepsilon_{eq}}\Lambda_3(\mu)d\mu - \frac{fa}{\sqrt{s - a}}.\tag{33}$$

Here $\Lambda_3(\varepsilon_{eq})$ denotes the function $\Lambda(\varepsilon_{eq})$ when $\beta = -1$ in (23). The value of ε_{eq} at $\zeta = \zeta_n$ is determined from (12) and (15) as

$$\varepsilon_n = -\frac{1}{\sqrt{3}}\ln\left[4\left(s - a\frac{ds}{da}\right)\right].\tag{34}$$

Then, the radial stress at $\zeta = \zeta_n$ is found from (33) in the form

$$\frac{\sigma_r^{(n)}}{\sigma_0} = \int_{\varepsilon_3}^{\varepsilon_n}\Lambda_3(\varepsilon_{eq})d\varepsilon_{eq} - \frac{fa}{\sqrt{s - a}}.\tag{35}$$

In region 2, Equation (20) becomes

$$\frac{\partial\sigma_r}{\sigma_0\partial\zeta} = \frac{a\Lambda_2(\zeta)}{\sqrt{3}(\zeta a + s)}.\tag{36}$$

Here the function $\Lambda_2(\zeta)$ is determined by replacing ε_{eq} and χ in $\Phi(\varepsilon_{eq}, \zeta_0\chi)$ with the right-hand sides of Equations (16) and (21), respectively. The radial stress must be continuous at $\zeta = \zeta_n$ and $\zeta = \zeta_{n0}$. Therefore, using (35), one can represent the solution of Equation (36) as

$$\frac{\sigma_r}{\sigma_0} = \frac{a}{\sqrt{3}}\int_{\zeta_n}^{\zeta}\frac{\Lambda_2(\mu)}{(\mu a + s)}d\mu + \int_{\varepsilon_3}^{\varepsilon_n}\Lambda_3(\varepsilon_{eq})d\varepsilon_{eq} - \frac{fa}{\sqrt{s - a}}.\tag{37}$$

Equations (30) and (37) combine to give

$$\int_{\varepsilon_1}^{\varepsilon_0}\Lambda_1(\varepsilon_{eq})d\varepsilon_{eq} - \frac{a}{\sqrt{3}}\int_{\zeta_n}^{\zeta_{n0}}\frac{\Lambda_2(\zeta)}{(\zeta a + s)}d\zeta - \int_{\varepsilon_3}^{\varepsilon_n}\Lambda_3(\varepsilon_{eq})d\varepsilon_{eq} + \frac{fa}{\sqrt{s - a}} = 0.\tag{38}$$

This equation connects a, s, and ds/da. Therefore, it can be considered an ordinary differential equation to find s as a function of a. Its solution must satisfy the condition (6).

Once Equation (38) has been solved, the distribution of the radial stress in all three regions is readily determined from (28), (33), and (37). Equations (28) and (33) should be used in conjunction with Equations (14) and (15), respectively. The distribution of the circumferential stress is found from (18). The bending moment is given by

$$M = \int_{r_{CD}}^{r_{AB}} \left(\sigma_\theta - \frac{F}{h} \right) r \, dr. \tag{39}$$

Here h is the current thickness of the sheet, $h = r_{AB} - r_{CD}$ (Figure 1). Using (8), one can rewrite (39) as

$$m = \frac{2\sqrt{3}M}{\sigma_0 H^2} = \frac{\sqrt{3}}{a} \int_{-1}^{0} \left(\frac{\sigma_\theta}{\sigma_0} - f\frac{H}{h} \right) d\zeta. \tag{40}$$

Note that $m = 1$ in the case of rigid perfectly plastic material if $f = 0$ [28].

5. Numerical Method

It has been noted that Equation (38) is an ordinary differential equation. Several well-established methods are available for solving such equations numerically (see, for example, [29]). However, the form of the differential equation in Equation (38) is non-standard. Therefore, a special numerical method should be developed to solve this equation. The method below is a modification of the method developed in [30] for solving a system of partial differential equations.

Equation (38), together with Equation (6), constitutes the initial value problem. However, a difficulty is that the equation provides no information at the initial instant. In particular, each term of this equation vanishes at $a = 0$ since $\varepsilon_{eq} = 0$ everywhere. It is a consequence of rigid plasticity. Such difficulty does not arise in elastic/plastic solutions (see, for example, [31]). Below, the difficulty noted is resolved by finding the state of stress at the initial instant without using Equation (38) and developing a general iterative procedure.

Assume that $s = s_i$ and $ds/da = q_i$ at $a = a_i$, and s_i and q_i are known. Let Δa be a small increment of a. Then, $a_{i+1} = a_i + \Delta a$, and the value of s at $a = a_{i+1}$ can be approximated as

$$s_{i+1} = s_i + \frac{1}{2}(q_i + q_{i+1})\Delta a. \tag{41}$$

Here q_{i+1} is the value of the derivative ds/da at $a = a_{i+1}$, and its value is unknown. Using (27), (29), (31), (34), and (41), one can express the limits of integration involved in (38) in terms of q_{i+1} and known quantities. Thus Equation (38) determines q_{i+1}. The procedure above can be repeated for $a = a_{i+2} = a_{i+1} + \Delta a$ and subsequent values of a.

It remains to find s and ds/da at $a = 0$ to start the iterative procedure above. The required value of s is given by (6). The initial cross-section of the sheet is shown in Figure 1a. The through-thickness stress vanishes at this instant. The equivalent strain vanishes everywhere. Equation (21) becomes

$$\chi = \begin{cases} \dfrac{4\gamma(\zeta+q_0)}{\sqrt{3}} & \text{in the range } -q_0 \leq \zeta \leq 0 \\[2ex] -\dfrac{4\gamma(\zeta+q_0)}{\sqrt{3}} & \text{in the range } -1 \leq \zeta \leq -q_0 \end{cases}. \tag{42}$$

Here q_0 is the value of the derivative ds/da at $a = 0$. Equations (2) and (42) combine to give

$$\frac{\sigma_\eta}{\sigma_0} = \begin{cases} \dfrac{2}{\sqrt{3}}\Phi\left[0, \dfrac{4\zeta_0\gamma(\zeta+q_0)}{\sqrt{3}}\right] & \text{in the range } -q_0 \leq \zeta \leq 0 \\[2ex] -\dfrac{2}{\sqrt{3}}\Phi\left[0, -\dfrac{4\gamma(\zeta+q_0)}{\sqrt{3}}\right] & \text{in the range } -1 \leq \zeta \leq -q_0 \end{cases}. \tag{43}$$

The tensile force at the initial instant is determined as

$$F = H \int_0^1 \sigma_\eta d\zeta.$$

(44)

Substituting (43) into (44) gives

$$f = \frac{2}{\sqrt{3}} \int_{-q_0}^{0} \Phi\left[0, \frac{4\xi_0\gamma(\zeta + q_0)}{\sqrt{3}}\right] d\zeta - \frac{2}{\sqrt{3}} \int_{-1}^{-q_0} \Phi\left[0, -\frac{4\xi_0\gamma(\zeta + q_0)}{\sqrt{3}}\right] d\zeta.$$

(45)

This equation determines q_0. In particular, $q_0 = 1/2$ if $f = 0$ (pure bending). The dimensionless bending moment can be found using (40), where σ_θ should be replaced with σ_η.

6. Numerical Example

The numerical results below are for pure bending. Several functions $\Phi(\varepsilon_{eq}, \xi_{eq})$ are available in the literature [18–24]. Since bending of strain-hardening materials, including quantitative results, has been investigated in many papers, the numerical example below emphasizes the effect of loading speed on process parameters.

In most cases, the function $\Phi(\varepsilon_{eq}, \xi_{eq})$ is represented as a product of two functions. One of these functions depends on the equivalent strain and the other on the equivalent strain rate. The illustrative example in this section assumes that $\Phi(\varepsilon_{eq}, \xi_{eq})$ is the product of Swift's strain-hardening law and Herschel–Bulkley's viscoplastic law. Then,

$$\Phi(\varepsilon_{eq}, \xi_{eq}) = (1 + \lambda\varepsilon_{eq})^\nu \left[1 + \left(\frac{\xi_{eq}}{\xi_0}\right)^n\right].$$

(46)

In all calculations, $\lambda = 0.73$, $\nu = 0.17$, and $n = 0.2$. The characteristic strain rate is involved in the solution only through γ. In particular, one can rewrite (46) as

$$\Phi(\varepsilon_{eq}, \chi\xi_0) = (1 + \lambda\varepsilon_{eq})^\nu (1 + \chi^n).$$

(47)

The numerical solution has been found using the procedure described in the previous section. It demonstrates the effect of γ on the through-thickness distribution of stresses and the bending moment. Figures 2 and 3 depict the through-thickness distributions of the radial and circumferential stresses, respectively, at several process stages identified by the inner radius of the sheet. These distributions correspond to $\gamma = 0.1$. In these figures,

$$Z = \frac{r - r_{CD}}{H}.$$

(48)

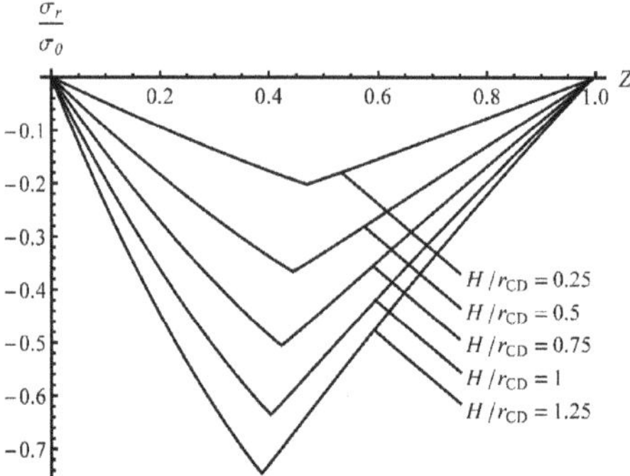

Figure 2. Through-thickness distribution of the radial stress at $\gamma = 0.1$ and several process stages.

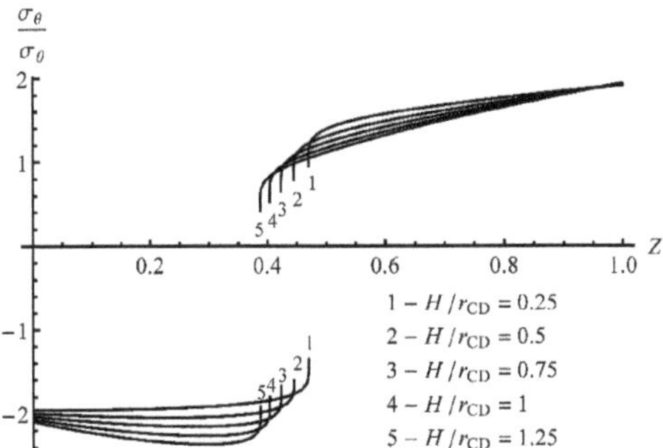

Figure 3. Through-thickness distribution of the circumferential stress at $\gamma = 0.1$ and several process stages.

Thus $Z = 0$ corresponds to the inner radius of the sheet and $Z = h/H$ to its outer radius. Using (8) and (9), one transforms Equation (48) to

$$Z = \sqrt{\frac{\zeta}{a} + \frac{s}{a^2}} - \sqrt{\frac{s}{a^2} - \frac{1}{a}}. \qquad (49)$$

At the final stage of the process, the inner radius of the sheet is rather small (smaller than the initial thickness of the sheet). Figures 4 and 5 show the same distributions at $\gamma = 1$. The circumferential stress is discontinuous at the neutral line. Figures 3 and 5 show that the neutral line moves to the inner surface of the sheet as the deformation proceeds. Thus the assumption made in Section 3 has been confirmed. The magnitude of the radial stress is quite high near the neutral line, except at the very beginning of the process (see Table 1). Therefore, the assumption used in simplified solutions [32,33] that the radial stress is negligible is not valid. The rightmost points of the curves in Figures 2–5 correspond to $Z = h/H$. Therefore, it is seen from these figures that the change in the thickness is negligible.

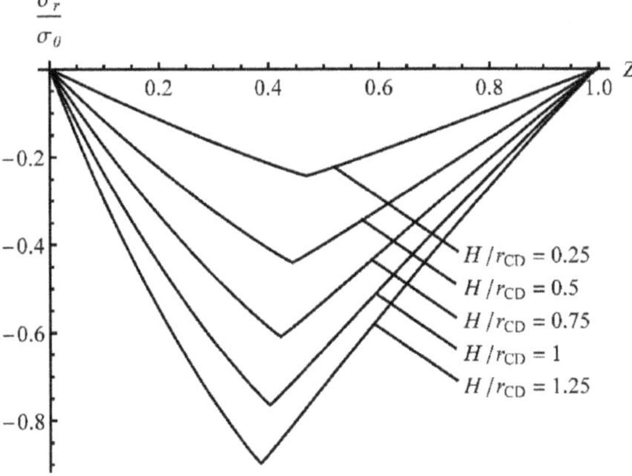

Figure 4. Through-thickness distribution of the radial stress at $\gamma = 1$ and several process stages.

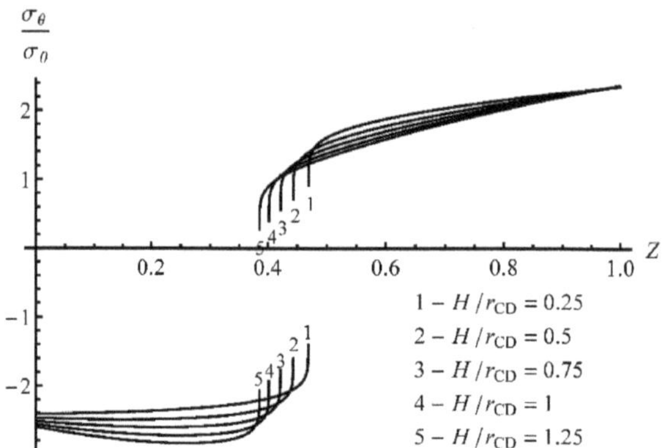

Figure 5. Through-thickness distribution of the circumferential stress at $\gamma = 1$ and several process stages.

Table 1. Radial stress at the neutral line.

H/r_{CD}	0.25	0.5	0.75	1	1.25
			$\gamma = 0.1$		
σ_r/σ_0	−0.2	−0.38	−0.51	−0.65	−0.74
			$\gamma = 1$		
σ_r/σ_0	−0.25	−0.45	−0.6	−0.75	−0.9

Figure 6 demonstrates the effect of γ on the dimensionless bending moment. The broken line corresponds to the rate-independent material model (i.e., the material obeys Swift's strain-hardening law). The bending moment increases as γ increases, which is in agreement with physical expectations. The increase in the bending moment is sharper for smaller values of γ. It is worthy of note that the bending moment is nearly constant for a given value of γ.

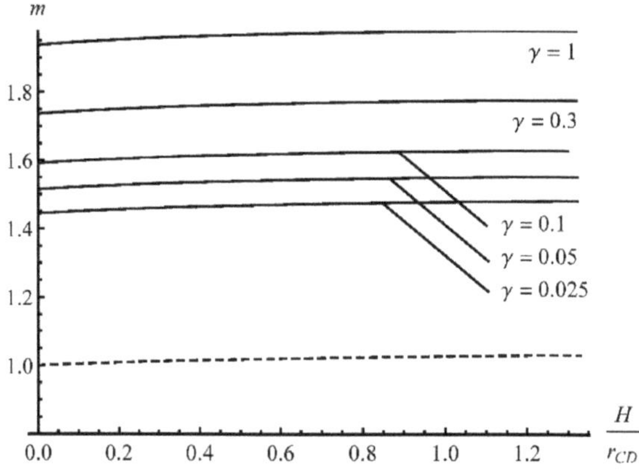

Figure 6. Variation of the dimensionless bending moment as the deformation proceeds at several values of γ. The broken line corresponds to the rate-independent material.

7. Conclusions

An accurate rigid plastic solution for a general strain-hardening viscoplastic material model has been found. The solution describes the plane strain process of bending under tension due to large strains. The elastic portion of the strain tensor has been neglected. However, the effect of elasticity on the accuracy of solutions is negligible except at the beginning of the process [34].

The numerical solution reduces to an ordinary differential equation in a non-standard form. A special numerical procedure is required to solve this equation. The procedure is described in Section 5. It is used to determine the stress field and the bending moment in the pure bending of a sheet made of a material obeying Equation (46). This numerical solution is presented and discussed in detail in Section 6.

The solution presented may be considered a starting point for designing experiments to identify constitutive equations. It is advantageous in this respect that the general solution is not restricted to a specific function or even a class of functions but is valid for an arbitrary dependence of the yield stress on the equivalent strain and the equivalent strain rate.

The solution can be used as a benchmark problem for verifying numerical codes. This necessary step applies to different constitutive equations, including equations used in metal forming applications [35–37].

Author Contributions: Conceptualization and writing, S.A.; formal analysis and writing E.L. All authors have read and agreed to the published version of the manuscript.

Funding: This research was made possible by grant 20-69-46070 from the Russian Science Foundation.

Institutional Review Board Statement: Not applicable.

Informed Consent Statement: Not applicable.

Data Availability Statement: Not applicable.

Conflicts of Interest: The authors declare no conflict of interest.

References

1. Mayville, R.A.; Finnie, I. Uniaxial stress-strain curves from a bending test. *Exp. Mech.* **1982**, *22*, 197–201. [CrossRef]
2. Oldroyd, P.W.J. Tension-compression stress-strain curves from bending tests. *J. Strain Anal.* **1971**, *6*, 286–292. [CrossRef]
3. Laws, V. Derivation of the tensile stress-strain curve from bending data. *J. Mater. Sci.* **1981**, *16*, 1299–1304. [CrossRef]
4. Kato, H.; Tottori, Y.; Sasaki, K. Four-point bending test of determining stress-strain curves asymmetric between tension and compression. *Exp. Mech.* **2014**, *54*, 489–492. [CrossRef]
5. Solyaev, Y.; Lurie, S.; Prokudin, O.; Antipov, V.; Rabinskiy, L.; Serebrennikova, N.; Dobryanskiy, V. Elasto-plastic behavior and failure of thick GLARE laminates under bending loading. *Compos. Part B Eng.* **2020**, *200*, 108302. [CrossRef]
6. Bernardo, L.; Nepomuceno, M.; Pinto, H. Neutral axis depth versus ductility and plastic rotation capacity on bending in lightweight-aggregate concrete beams. *Materials* **2019**, *12*, 3479. [CrossRef]
7. Alves, M.M.; Abreu, L.M.; Pereira, A.B.; De Morais, A.B. Determination of bilinear elastic-plastic models for an acrylic adhesive using the adhesively bonded 3-point bending specimen. *J. Adhes.* **2021**, *97*, 553–568. [CrossRef]
8. Rupp, P.; Elsner, P.; Weidenmann, K.A. Failure mode maps for four-point-bending of hybrid sandwich structures with carbon fiber reinforced plastic face sheets and aluminum foam cores manufactured by a polyurethane spraying process. *J. Sandw. Struct. Mater.* **2019**, *21*, 2654–2679. [CrossRef]
9. Dadras, P.; Majlessi, S.A. Plastic bending of work hardening materials. *ASME. J. Eng. Ind.* **1982**, *104*, 224–230. [CrossRef]
10. Dadras, P. Plane strain elastic–plastic bending of a strain-hardening curved beam. *Int. J. Mech. Sci.* **2001**, *43*, 39–56. [CrossRef]
11. Heller, B.; Kleiner, M. Semi-analytical process modelling and simulation of air bending. *J. Strain Anal. Eng. Des.* **2006**, *41*, 57–80. [CrossRef]
12. Ahn, K. Plastic bending of sheet metal with tension/compression asymmetry. *Int. J. Solids Struct.* **2020**, *204–205*, 65–80. [CrossRef]
13. Alexandrov, S.; Kim, J.-H.; Chung, K.; Kang, T.-J. An alternative approach to analysis of plane-strain pure bending at large strains. *J. Strain Anal. Eng. Des.* **2006**, *41*, 397–410. [CrossRef]
14. Alexandrov, S.; Lyamina, E.; Hwang, Y.-M. Plastic bending at large strain: A review. *Processes* **2021**, *9*, 406. [CrossRef]
15. Parsa, M.; Rad, M.; Shahhosseini, M.R.; Shahhosseini, M. Simulation of windscreen bending using viscoplastic formulation. *J. Mater. Process. Technol.* **2005**, *170*, 298–303. [CrossRef]
16. Alvarez González, C.M.; Okabe, T. Application of multiplicative viscoplastic model to deformation analysis of creep bending test of structural steel. *J. Struct. Constr. Eng.* **2018**, *83*, 1343–1351. [CrossRef]
17. Saito, N.; Fukahori, M.; Minote, T.; Funakawa, Y.; Hisano, D.; Hamasaki, H.; Yoshida, F. Elasto-viscoplastic behavior of 980 MPa nano-precipitation strengthened steel sheet at elevated temperatures and springback in warm bending. *Int. J. Mech. Sci.* **2018**, *146–147*, 571–582. [CrossRef]

18. Perrone, N. A mathematically tractable model of strain-hardening, rate-sensitive plastic flow. *ASME. J. Appl. Mech.* **1966**, *33*, 210–211. [CrossRef]

19. Shang, H.; Wu, P.; Lou, Y. Strain hardening of AA5182-O considering strain rate and temperature effect. In *Forming the Future. The Minerals, Metals & Materials Series*; Daehn, G., Cao, J., Kinsey, B., Tekkaya, E., Vivek, A., Yoshida, Y., Eds.; Springer: Cham, Switzerland, 2021; pp. 657–665. [CrossRef]

20. De Angelis, F.; Cancellara, D.; Grassia, L.; D'Amore, A. The influence of loading rates on hardening effects in elasto/viscoplastic strain-hardening materials. *Mech. Time-Depend. Mater.* **2018**, *22*, 533–551. [CrossRef]

21. Lomakin, E.; Fedulov, B.; Fedorenko, A. Strain rate influence on hardening and damage characteristics of composite materials. *Acta Mech.* **2021**, *232*, 1875–1887. [CrossRef]

22. Li, Z.; Wang, J.; Yang, H.; Liu, J.; Ji, C. A modified Johnson–Cook constitutive model for characterizing the hardening behavior of typical magnesium alloys under tension at different strain rates: Experiment and simulation. *J. Mater. Eng. Perform.* **2020**, *29*, 8319–8330. [CrossRef]

23. Heydarian, M.; Shafiei, E.; Ostovan, F. A phenomenological model utilizing Voce's strain-hardening model describing high strain rate behavior of alloy 800H during hot working. *J. Strain Anal. Eng. Des.* **2020**, *55*, 271–276. [CrossRef]

24. Ghazani, M.S.; Eghbali, B. Strain hardening behavior, strain rate sensitivity and hot deformation maps of AISI 321 austenitic stainless steel. *Int. J. Miner. Metall. Mater.* **2021**. [CrossRef]

25. Durban, D.; Baruch, M. Analysis of an elasto-plastic thick walled sphere loaded by internal and external pressure. *Int. J. Non-Linear Mech.* **1977**, *12*, 9–21. [CrossRef]

26. Alexandrov, S.; Hwang, Y.-M. Plane strain bending with isotropic strain hardening at large strains. *Trans. ASME J. Appl. Mech.* **2010**, *77*, 064502. [CrossRef]

27. Alexandrov, S.; Pirumov, A.; Jeng, Y.-R. Expansion/contraction of a spherical elastic/plastic shell revisited. *Cont. Mech. Therm.* **2015**, *27*, 483–494. [CrossRef]

28. Hill, R. *The Mathematical Theory of Plasticity*; Clarendon Press: Oxford, UK, 1950.

29. Atkinson, K.; Han, W.; Stewart, D. *Numerical Solution of Ordinary Differential Equations*; John Wiley & Sons: Hoboken, NJ, USA, 2009.

30. Alexandrov, S.; Gelin, J.-C. Plane strain pure bending of sheets with damage evolution at large strains. *Int. J. Solids Struct.* **2011**, *48*, 1637–1643. [CrossRef]

31. Alexandrov, S.; Manabe, K.; Furushima, T. A general analytic solution for plane strain bending under tension for strain-hardening material at large strains. *Arch. Appl. Mech.* **2011**, *81*, 1935–1952. [CrossRef]

32. Yuen, W.Y.D. A generalised solution for the prediction of springback in laminated strip. *J. Mater. Process. Technol.* **1996**, *61*, 254–264. [CrossRef]

33. Kagzi, S.A.; Gandhi, A.H.; Dave, H.K.; Raval, H.K. An analytical model for bending and springback of bimetallic sheets. *Mech. Adv. Mater. Struct.* **2016**, *23*, 80–88. [CrossRef]

34. Alexandrov, S.; Lyamina, E.; Hwang, Y.-M. Finite pure plane strain bending of inhomogeneous anisotropic sheets. *Symmetry* **2021**, *13*, 145. [CrossRef]

35. Roberts, S.M.; Hall, F.; Van Bael, A.; Hartley, P.; Pillinger, I.; Sturgess, E.N.; Van Houtte, P.; Aernoudt, E. Benchmark tests for 3-D, elasto-plastic, finite-element codes for the modelling of metal forming processes. *J. Mater. Process. Technol.* **1992**, *34*, 61–68. [CrossRef]

36. Abali, B.E.; Reich, F.A. Verification of deforming polarized structure computation by using a closed-form solution. *Contin. Mech. Thermodyn.* **2020**, *32*, 693–708. [CrossRef]

37. Yang, H.; Timofeev, D.; Abali, B.E.; Li, B.; Muller, W.H. Verification of strain gradient elasticity computation by analytical solutions. *ZAMM* **2021**, *101*, e202100023. [CrossRef]

Article

Tube Expansion by Single Point Incremental Forming: An Experimental and Numerical Investigation

Carlos Suntaxi [1,2], Gabriel Centeno [2,*], M. Beatriz Silva [3], Carpóforo Vallellano [2] and Paulo A. F. Martins [3]

[1] Departamento de Ingeniería Mecánica, Facultad de Ingeniería Mecánica, Escuela Politécnica Nacional, Ladrón de Guevara E11-253, Quito P.O. Box 17-01-2759, Ecuador; segundo.suntaxi@epn.edu.ec

[2] Department of Mechanical and Manufacturing Engineering, School of Engineering, University of Seville, Camino de los Descubrimientos s/n, 41092 Seville, Spain; carpofor@us.es

[3] IDMEC, Instituto Superior Técnico, Universidade de Lisboa, Av. Rovisco Pais, 1049-001 Lisboa, Portugal; beatriz.silva@tecnico.ulisboa.pt (M.B.S.); pmartins@tecnico.ulisboa.pt (P.A.F.M.)

* Correspondence: gaceba@us.es; Tel.: +34-954-485-965

Abstract: In this paper, we revisit the formability of tube expansion by single point incremental forming to account for the material strain hardening and the non-proportional loading paths that were not taken into consideration in a previously published analytical model of the process built upon a rigid perfectly plastic material. The objective is to provide a new insight on the reason why the critical strains at failure of tube expansion by single point incremental forming are far superior to those of conventional tube expansion by rigid tapered conical punches. For this purpose, we replaced the stress triaxiality ratio that is responsible for the accumulation of damage and cracking by tension in monotonic, proportional loading paths, by integral forms of the stress triaxiality ratio that are more adequate for the non-proportional paths resulting from the loading and unloading cycles of incremental tube expansion. Experimental and numerical simulation results plotted in the effective strain vs. stress triaxiality space confirm the validity of the new damage accumulation approach for handling the non-proportional loading paths that oscillate cyclically from shearing to biaxial stretching, as the single point hemispherical tool approaches, contacts and moves away from a specific location of the incrementally expanded tube surface.

Keywords: single point incremental forming; tube expansion; formability; fracture; stress-triaxiality

Citation: Suntaxi, C.; Centeno, G.; Silva, M.B.; Vallellano, C.; Martins, P.A.F. Tube Expansion by Single Point Incremental Forming: An Experimental and Numerical Investigation. *Metals* **2021**, *11*, 1481. https://doi.org/10.3390/met11091481

Academic Editor: Zhengyi Jiang

Received: 18 August 2021
Accepted: 14 September 2021
Published: 17 September 2021

Publisher's Note: MDPI stays neutral with regard to jurisdictional claims in published maps and institutional affiliations.

1. Introduction

The route for characterizing the sheet formability limits started in the late 1960s when Keeler [1] and Goodwin [2] developed the circle grid analysis (CGA) technique for determining the in-plane strains on the surface of sheet metal formed parts. The use of principal strain space to plot these strains and to identify their critical values at the onset of failure by Embury and Duncan [3] in the early 1980s paved the way to what they called "formability maps", which are nowadays known as the forming limit diagrams (FLDs) [4].

A typical FLD for a sheet metal forming material is built on three different types of failure limit curves [5]: (i) the forming limit curve (FLC) corresponding to failure by necking, (ii) the fracture forming limit lines corresponding to failure by cracking and (iii) the wrinkling limit curve (WLC) delimiting the onset of wrinkling in the lower left-hand of the second quadrant. In sheet metal forming, there are two fracture forming lines corresponding to crack opening by tension (mode I of fracture mechanics, hereafter referred to as FFL) and crack opening by in-plane shear (mode II of fracture mechanics, hereafter referred to as SFFL) [6]. The experimental determination of the FFLs and SFFLs was comprehensively explained by the authors in previous publications [4,5], who also described the different methods and procedures to obtain the FLCs.

The route for establishing the formability limits of tube materials starts with the determination of the onset of necking (FLC) by means of tube of hydroforming [7,8]. No methodologies for characterizing the crack opening modes and determining the fracture forming lines were proposed until 2016, when Centeno et al. [9] utilized CGA to plot the FLC and the FFL corresponding to tube cracking by tension.

Subsequent research work combining numerical methods, digital image correlation (DIC) and making use of a broader range of tube forming processes comprising expansion [10], inversion [11] and bulging [12] allowed obtaining the FLC and the FFL of tube materials for a wider range of strain paths running from uniaxial tension up to equal biaxial stretching (e.g., from strain ratios $\beta = d\varepsilon_2/d\varepsilon_1$ ranging between $-1/2$ to 1). These efforts were recently complemented by the work of Magrinho et al. [13], who proposed an experimental procedure to determine the SFFL of tube materials (i.e., the fracture forming limit line corresponding to tube cracking by in-plane shear).

In view of the aforementioned work, recent developments in incremental tube expansion, reduction, wall grooving and hole flanging using a single point hemispherical tool by Wen et al. [14] and Movahedinia et al. [15] raise the question of whether their deformation mechanics and formability limits remain the same as those of conventional tube forming processes. The answer to this question was firstly addressed by Cristino et al. [16], who presented an analytical model based on membrane analysis for tube expansion by single point incremental forming (hereafter referred to as incremental tube expansion). The model reveals the main differences between conventional and incremental tube expansion in terms of stress/strain states and damage accumulation to explain the greater formability of incremental tube expansion compared to that of conventional tube expansion with a rigid tapered conical punch.

The analytical model of Cristino et al. [16] is based on a rigid, perfectly plastic tube material and assumes near-proportional (equal biaxial stretching) experimental strain loading paths in principal strain space to facilitate algebraic treatment.

Under these circumstances, it is important to revisit the accumulation of damage by means of a numerical simulation model capable of accounting for material strain hardening and for the loading/unloading cycles of incremental tube expansion. In this paper, we provide a novel perspective on the formability and failure of incremental tube forming processes subjected to non-proportional loading. We analyze different methodologies to account for stress triaxiality and accumulation of damage, and discuss if the FFLs of tube materials determined by means of conventional tube forming processes subjected to near proportional loading paths are still valid for incremental tube expansion characterized by non-proportional loading paths that oscillate cyclically from shearing to biaxial stretching, as the single point hemispherical tool approaches, contacts and moves away from a specific location of the plastically deformed tube surface.

Experimental and numerical simulation results plotted in the space of effective strain vs. stress triaxiality [17] give support to the discussion, which is of paramount importance to infer about the FFLs of tubes being material properties, in contrast to their FLCs, which are dependent on the applied loading paths.

2. Methods and Procedures

The investigation was carried out in AA6063-T6 extruded aluminum tubes with an outer radius $r_0 = 20$ mm and a wall thickness $t_0 = 2$ mm. The first part of this section summarizes the methods and procedures that were utilized to determine the material flow curve and the formability limits by necking (FLC) and by fracture under tension (FFL) using conventional tube forming processes. The data provided in the figures were retrieved from previous publications of the authors [9–12].

In the second part of this section, we present the experimental testing conditions of incremental tube expansion, describe the methodology that was used to determine the strain paths using circle grid analysis (CGA), provide an analytical framework to transform the formability limits from principal strain space into the effective strain vs. stress-triaxiality space and summarize the numerical modelling conditions utilized in finite element analysis.

2.1. Flow Curve

The flow curve of the AA6063-T6 tubes is shown in Figure 1 and was obtained by merging the stress–strain evolutions that were previously obtained by the authors using tensile and stack compression tests [9]. Tensile tests were carried out in specimens machined out from the tube longitudinal direction and provided the material stress response for values of effective strain below 0.23 (refer to the vertical dashed line). Stack compression tests were performed in cylindrical specimens that were assembled by pilling up disks that were also machined out from the supplied tubes and allowed characterizing the strain hardening behavior of the tube material for the remaining values of effective strain.

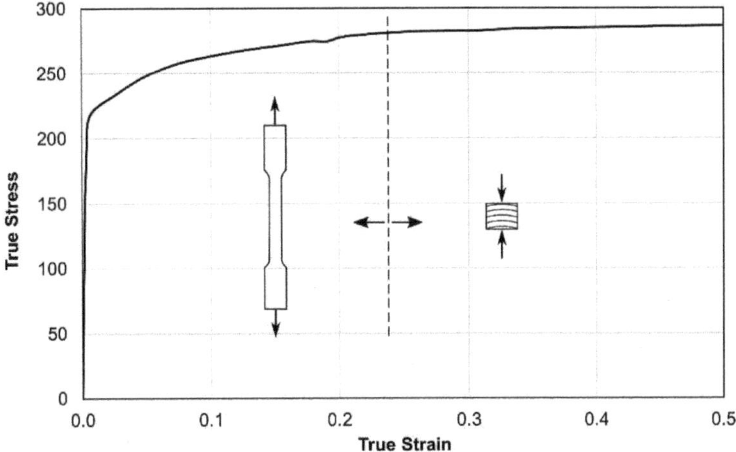

Figure 1. Flow curve of the aluminum AA6063T6 tubes (adapted from [9]).

2.2. Formability Limits by Necking and Fracture

Figure 2 shows the formability limits of the aluminum AA6063-T6 tubes by necking (FLC) and by fracture under tension (FFL) in principal strain space. Determination of the FLC required measuring the strain paths of conventional tube expansion, inversion and bulging by means of digital image correlation (DIC) and combining these results with time-dependent and force-dependent methodologies that were specifically developed by the authors for tube materials [11,12]. Determination of the FFL required measuring the wall thickness of the tube cracked regions by optical microscopy (D software version 5.11.01, NIS-Elements, Tokyo, Japan)to obtain the "gauge length" strains at fracture. Information about the different tests, methods and procedures that were used by the authors to determine the FLCs and FFLs of tube materials is given in Magrinho et al. [12].

Figure 2 also includes the strain loading path obtained in conventional tube expansion with a rigid tapered conical punch having a semi-angle of 15°, which was previously obtained by the authors [12] and will be used for reference purposes throughout this paper. As seen, the in-plane strains of conventional tube expansion at the onset of necking $(\varepsilon_{1n}, \varepsilon_{2n}) = (-0.25, 0.41)$ are very close to the FLC, and the in-plane fracture strains $(\varepsilon_{1f}, \varepsilon_{2f}) = (-0.25, 0.71)$ are exactly on top of the FFL.

The formability limits shown in Figure 2 can alternatively be plotted in the effective strain vs. stress triaxiality space (Figure 3). The transformation of the formability limits from principal strain space into this other space can be carried out analytically by assuming linear, proportional strain paths under plane stress loading conditions ($\sigma_t = \sigma_3 \approx 0$). Plane stress loading conditions are commonly assumed in the analytical modelling of sheet and thin-wall tube forming [12,18].

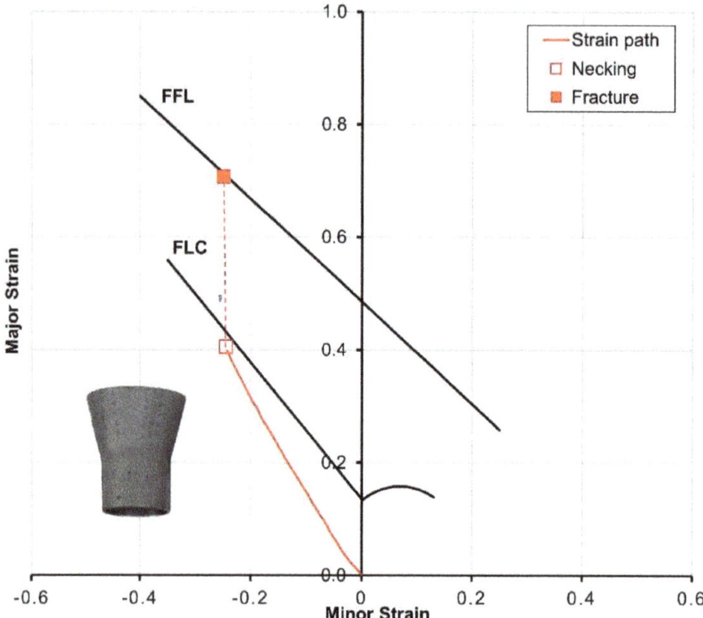

Figure 2. Forming limit curve (FLC) and fracture forming limit (FFL) line of the aluminum AA6063-T6 tube in principal strain space. The red line represents the experimental strain loading path of conventional tube expansion with a rigid tapered conical punch having a semi-angle of 15° (adapted from [12]).

For this purpose, let us consider, for example, the tube material to be isotropic and to follow the von Mises yield criterion, so that its effective stress $\bar{\sigma}$ and effective strain $d\bar{\varepsilon}$ are given by:

$$\bar{\sigma} = \sqrt{\sigma_1^2 - \sigma_1\sigma_2 + \sigma_2^2} \tag{1}$$

$$d\bar{\varepsilon} = \frac{2}{\sqrt{3}}\sqrt{d\varepsilon_1^2 + d\varepsilon_1 d\varepsilon_2 + d\varepsilon_2^2} \tag{2}$$

Then, applying the Levy–Mises constitutive equations, one obtains the following relation between the stress triaxiality ratio $\eta = \sigma_m/\bar{\sigma}$ and the slope $\beta = d\varepsilon_2/d\varepsilon_1$ of the strain path [9]:

$$\eta = \frac{1+\beta}{\sqrt{3}\,\sqrt{1+\beta+\beta^2}} \tag{3}$$

The above equation together with the following modified version of Equation (2) to include the slope β in the effective strain,

$$\bar{\varepsilon} = \frac{2}{\sqrt{3}}\sqrt{1+\beta+\beta^2}\,\varepsilon_1\,, \tag{4}$$

allows accomplishing the above-mentioned transformation of the FLC from principal strain space into the effective strain vs. stress triaxiality space (Figure 3).

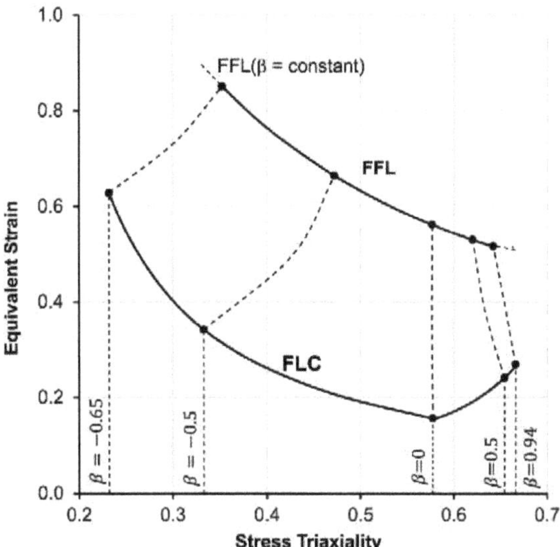

Figure 3. Forming limit curve (FLC) and fracture forming limit (FFL) line of the aluminum AA6063-T6 tube in the effective strain vs. stress triaxiality space, obtained from analytical transformation assuming material isotropy, linear strain paths and plane stress loading conditions.

The transformation of the FFL from principal strain space into the effective strain vs. stress triaxiality space requires consideration of the experimentally observed strain path deviation towards plane strain deformation conditions at the onset of necking (FLC); see, for instance, Martinez-Donaire et al. [19]. In case of the effective strain $\bar{\varepsilon}$, this is realized by modifying Equation (4) to account for the two piecewise linear strain paths involving the initial path (up to necking) with a given slope β and the final path (from necking to fracture) with a slope $\beta = 0$ resulting from strain localization in the tube material:

$$\bar{\varepsilon}_f = \int_0^{\bar{\varepsilon}_n} d\bar{\varepsilon} + \int_{\bar{\varepsilon}_n}^{\bar{\varepsilon}_f} d\bar{\varepsilon} = \frac{2}{\sqrt{3}} \left[\varepsilon_{1f} + \left(\sqrt{1+\beta+\beta^2} - 1 \right) \left(\varepsilon_{2f}/\beta \right) \right] \tag{5}$$

In the above equation, ε_{1f} and ε_{2f} are the major and minor in-plane strains at fracture, and $\bar{\varepsilon}_f$ is the effective strain at fracture.

In case of the stress triaxiality η, the transformation is carried out in accordance with Martinez-Donaire et al. [19], who introduced an integral form $\bar{\eta}_f$, named average-stress triaxiality at fracture, that accounts for stress triaxiality in an average sense over the two piecewise linear strain paths:

$$\bar{\eta}_f = \frac{1}{\bar{\varepsilon}_f} \int_0^{\bar{\varepsilon}_f} \frac{\sigma_m}{\bar{\sigma}} d\bar{\varepsilon} = \frac{1}{\bar{\varepsilon}_f} \left(\int_0^{\bar{\varepsilon}_n} \frac{\sigma_m}{\bar{\sigma}} d\bar{\varepsilon} + \int_{\bar{\varepsilon}_n}^{\bar{\varepsilon}_f} \frac{\sigma_m}{\bar{\sigma}} d\bar{\varepsilon} \right) = \frac{\sqrt{3}}{3} \left[\frac{\varepsilon_{1f} + \varepsilon_{2f}}{\varepsilon_{1f} + \left(\sqrt{1+\beta+\beta^2} - 1 \right) \left(\varepsilon_{2f}/\beta \right)} \right] \tag{6}$$

The FLC and FFL resulting from the above-mentioned analytical transformation procedure are shown in Figure 3 and are slightly different from those obtained by Magrinho et al. [12] due to the following two main reasons. First, the authors made use of the von Mises yield criterion instead of the Hosford yield criterion that was utilized by Magrinho et al. [12]. Second, Magrinho et al. [12] transformed the FFL from principal strain space into the effective strain vs. stress triaxiality space by replacing the strains at fracture directly on Equations (3) and (4) instead of using the two piecewise linear strain path approach given by Equations (5) and (6), i.e., without considering the kink in the strain loading path from necking towards fracture.

To conclude, it is worth mentioning that the main reason why the Hosford yield criterion was not utilized in this work was due to its unavailability in the commercial finite element computer program utilized by the authors. Hill's 48 yield criterion was not considered as well because of the difficulty in obtaining the Lankford's coefficient at 45° in a tube, and because Cristino et al. [16] achieved good analytical estimates of material flow neglecting anisotropy.

2.3. Incremental Tube Expansion

The experiments in incremental tube expansion were performed in a Deckel Maho CNC machining center equipped with a single point hemispherical tool ($r_{tool} = 5$ mm) made from a cold working tool steel (120WV4-DIN) hardened and tempered to 60 HRc. The bottom tube end of the specimens was fixed to prevent sliding and rotation, and the tool path was programmed to perform a multi-stage incremental forming sequence consisting of an upward helical trajectory with a constant semi-angle $\Psi = 15°$ (Figure 4a). The pitch p between two consecutive stages was set to 2 mm and the initial tool depth at the beginning of the first stage was set to 2 mm with respect to the upper tube end (Figure 3b). Table 1 summarizes the main process parameters.

Three different tests were performed under the above-mentioned experimental conditions and a total of eight forming stages were needed for each test to observe of an incipient failure by fracture close to the plastically deformed tube end (Figure 3c), as is later discussed in the paper.

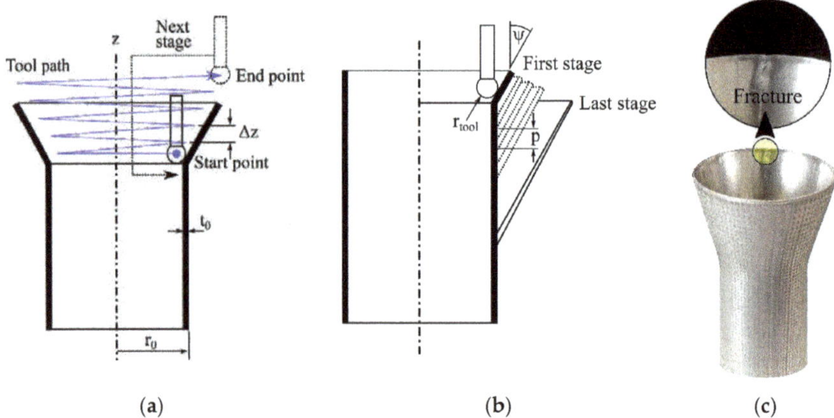

Figure 4. Schematic representation of incremental tube expansion showing (**a**) the single point hemispherical tool path and (**b**) the multi-stage forming strategy. A photograph of a specimen after eight forming stages is included in (**c**) together with a detail showing the final cracked surface.

Table 1. Main parameters used in the incremental tube expansion tests.

Parameter	Value
Tool radius r_{tool}	5 mm
Pitch p	2 mm
Step down Δz	0.2 mm
Semi-angle of inclination Ψ	15°
Feed rate	1000 mm/min

2.4. Strain Measurement Using Circle Grid Analysis

The in-plane strains of the incrementally expanded tubes were determined by CGA using the automatic measurement system ARGUS® v.6.2 by GOM™ equipped with a camera having a resolution of 1624 × 1236 pixels. For this purpose, the outer tube surfaces were electrochemically etched with a grid of circles with 0.75 mm of diameter and a distance between centers of 1.5 mm.

Measurement and classification of the deformed circles into different colors by ARGUS at the end of incremental tube expansion (Figure 5a) allowed determining the in-plane distribution of strains along the longitudinal direction from the undeformed lower tube region to the upper end of the plastically expanded tube surface (Figure 5b). The results for a typical longitudinal cross section marked with a black line in Figure 5a are given by the corresponding black line in principal strain space (Figure 5b).

(a)

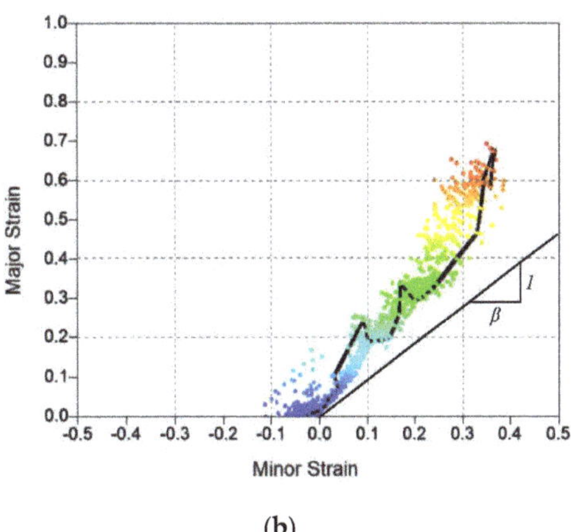

(b)

Figure 5. (a) Experimental determination of the in-plane strains by CGA using the automatic measuring system ARGUS®, and (b) representation of these strains in principal strain space.

2.5. Numerical Modelling

Numerical modelling of the conventional and incremental tube expansion processes was carried out with the commercial finite element computer program DEFORM™-3D. DEFORM™-3D was chosen due its capability to obtain a good agreement between numerical and experimental strains in incremental sheet forming processes [20,21].

The tube material was assumed as isotropic, elastic and plastic, and its initial geometry was discretized by means of solid tetrahedral elements. Tube material properties were taken from a previous work of the authors [9]. The tools were modelled as rigid (non-deformable) bodies and discretized by means of spatial triangular elements.

A penalty contact algorithm was utilized to model the interaction between the tools and the tube material.

Discretization of the tube and tool in case of conventional tube expansion with a rigid tapered conical punch took advantage of the rotational symmetry conditions of the process to create a simple three-dimensional model built upon an angular sector of 18° (1/20 of the full three-dimensional model). A total of 11,530 tetrahedral elements were utilized with an average side length of 1 mm and a reduced side length of 0.25 mm in the upper tube regions where mesh refinement was needed. Figure 6 shows the initial and final deformed meshes with the predicted contour of effective strain at the end of the process.

Typical CPU time to complete the numerical modelling of conventional tube expansion was approximately 3 min in a personal computer equipped with an Intel I7-4749 CPU (3.6 GHz) processor.

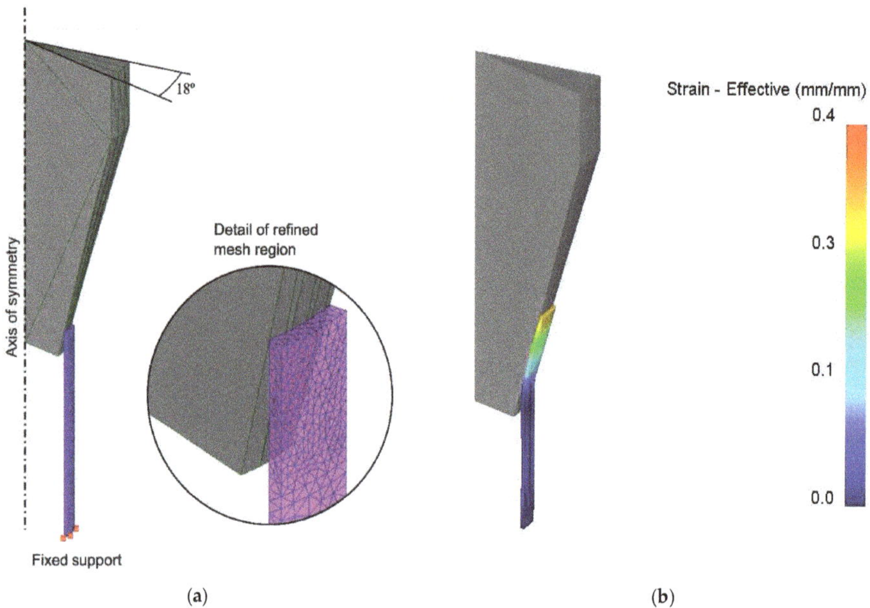

Figure 6. Finite element modeling of conventional tube expansion with a rigid tapered conical punch having a semi-angle of 15°. (**a**) Initial mesh with a detail of mesh refinement and (**b**) predicted distribution of effective strain at the end of the process.

Discretization of the tube material and of the single point hemispherical tool in case of incremental tube expansion required a full three-dimensional finite element model. The initial mesh consisted of 50,000 tetrahedral elements distributed along a finer mesh region at the upper tube end, which initially comes into contact with the tool, and a coarser mesh region for the remaining regions of the tube (Figure 7a). The final mesh at the end of the process (Figure 7b) consisted of approximately 120,000 tetrahedral elements due to several remeshings (based on critical element distortion) that were automatically performed to keep the numerical simulation from stopping because of excessive element distortion.

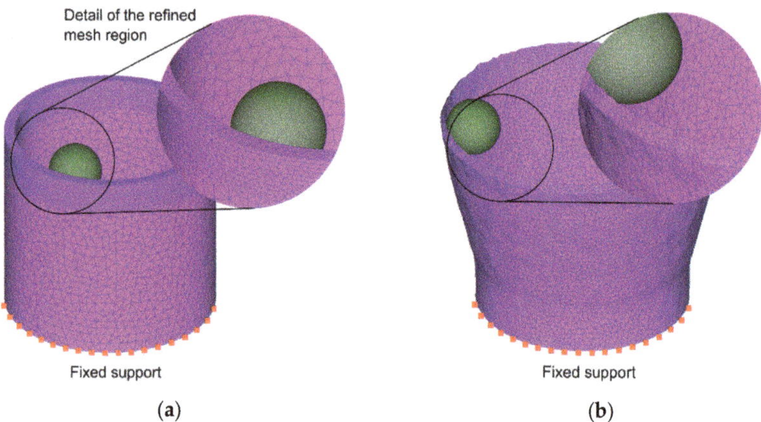

Figure 7. Finite element modeling of incremental tube expansion. (**a**) Initial mesh and (**b**) final mesh after eight forming stages.

The trajectory of the single point hemispherical tool was set identical to that utilized in the experiments and the total CPU time to complete the eight forming stages of the multi-stage incremental tube expansion process was approximately equal to 600 h. Cyclic strain loading paths of incremental tube forming cannot be obtained from experimental strain analysis; therefore, finite element predicted strain loading paths become the closest as one can be from the physical phenomenon, despite the long computational time.

3. Results and Discussion

This section is organized in two main parts. In the first part, we make use of the principal strain space and focus on material flow and failure by cracking in incremental tube expansion. Results obtained from conventional tube expansion are included for comparison purposes. In the second part, we discusses the application of different integral forms of stress triaxiality in the effective strain vs. stress triaxiality space to solve an apparent contradiction: on one hand, the in-plane strains of incremental tube expansion exceed the threshold admissible values of the FFL, which were previously determined by means conventional tube forming processes, and on the other hand, the FFL is a material property, and therefore, its threshold values cannot be surpassed and must be independent of any type of applied loading.

3.1. Material Flow and Cracking

Figure 8 shows the experimental in-plane strains along the longitudinal cross sections of the incremental tube expansion specimens in principal strain space. The results were obtained by ARGUS® (refer to the open triangular markers) for three test repetitions consisting of eight forming stages each. Numerical predictions obtained by finite element modelling with DEFORM™-3D are enclosed and confirm that incremental tube expansion subject the material to biaxial stretching conditions.

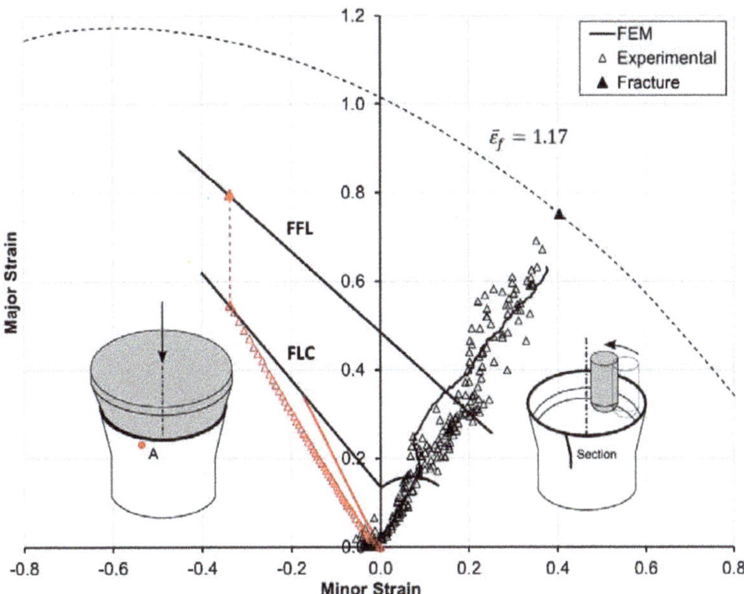

Figure 8. Experimental vs. finite element predicted in-plane strains for conventional tube expansion with a rigid tapered conical punch (red markers) and incremental tube expansion with a single point hemispherical tool after eight forming stages (black markers).

The in-plane strains at the onset of fracture by tension are represented by the solid black triangular marker and its determination involved measuring the tube wall thickness at the vicinity of the incipient cracking zone and calculating the through-thickness strain ε_{3f} at fracture to obtain the 'gauge length' strains $\left(\varepsilon_{1f}, \varepsilon_{2f}\right)$.

This alternative procedure was necessary because neither ARGUS® nor DEFORM™-3D could provide the in-plane strains at fracture. In fact, the application of circle grids with very small diameters to obtain the in-plane strains in the cracked region by means of ARGUS® is not feasible because it creates measurement problems and delivers values that cannot be considered fracture strains due to inhomogeneous material deformation around the cracks. Similar problems exist in finite element modelling with the use of very refined meshes in the regions where the cracks are likely to be triggered, plus the additional difficulty resulting from these results being sensitive to mesh size.

Under these circumstances, the authors had to measure the tube wall thickness in a NICON® SMZ800 optical microscope equipped with a NIS-Elements® D software version 5.11.01. Figure 9 shows a longitudinal cross-section detail after completion of the incremental tube expansion process with the corresponding evolution of thickness along the longitudinal direction (starting from the upper tube end). As seen, two different regions may be distinguished: (i) a first region (labeled as "I") located near the upper tube end that is characterized by a sharp decrease in wall thickness and (ii) a second region (labelled as "II"), in which the wall thickness progressively increases, as the distance to the upper tube end increases and approaches the undeformed region (not subjected to incremental expansion), along which the tube wall thickness $t_0 = 2$ mm remained unchanged.

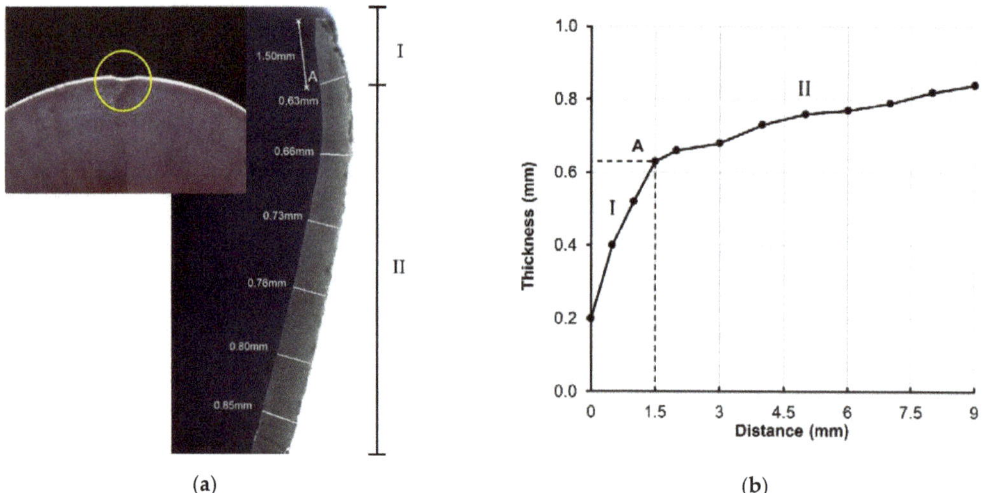

(a)   (b)

Figure 9. (**a**) Detail of a tube section after the incremental expansion and (**b**) evolution of the tube wall thickness with the longitudinal distance to the upper tube end.

Two main conclusions can be inferred from Figure 9: (i) failure by cracking is not preceded by necking and (ii) failure by cracking is related to a sharp decrease in the tube wall thickness in a small region "I" subjected to a great amount of straining. As seen, there is no localized thickness reduction in the detail of the tube section after the last forming stage of incremental expansion. This observation combined with the monotonic increase in the strain loading path up to fracture shown in Figure 8 (refer to the black triangular markers) allow concluding that failure occurs without previous necking.

A closer observation of the tube wall thickness within region "I" confirms the existence of micro-cracks along its length, as it was previously stated by Cristino et al. [16], and justifies the reason why the experimental determination of the "gauge length" strains

$\left(\varepsilon_{1f},\varepsilon_{2f}\right)$ at fracture was made at point "A" (Figure 9) located 1.5 mm away from the upper tube end, in the transition between regions "I" and "II".

The finite element predicted evolution of the in-plane strains at point "A" for each individual stage of the incremental tube expansion process is shown in Figure 10. As mentioned before, the "gauge length" strains $\left(\varepsilon_{1f},\varepsilon_{2f}\right)$ at fracture (refer to the black solid triangular marker) were not obtained by finite elements and their determination made use of the tube wall thickness value at point "A" (Figure 9b) to calculate the through-thickness strain ε_{3f} at fracture.

$$\varepsilon_{3f} = ln\frac{t_f}{t_0} = ln\frac{0.63}{2} = -1.16 \tag{7}$$

Then, assuming material incompressibility and the final slope β of the strain loading path to remain identical to that of the last piecewise linear path obtained by ARGUS® (Figure 8), it was possible to determine the 'gauge length' strains $\left(\varepsilon_{1f},\varepsilon_{2f}\right)$ at fracture, as follows:

$$\beta = \frac{\varepsilon_2}{\varepsilon_1} = 0.54 \quad \rightarrow \quad \varepsilon_{1f} = -\frac{\varepsilon_{3f}}{1+\beta} = 0.75 \quad , \quad \varepsilon_{2f} = \beta\varepsilon_{1f} = 0.40 \tag{8}$$

The corresponding effective strain at fracture $\bar{\varepsilon}_f = 1.17$ was obtained from Equation (4) and defines a dashed ellipse of constant effective strain values in principal strain space (refer to both Figures 8 and 9).

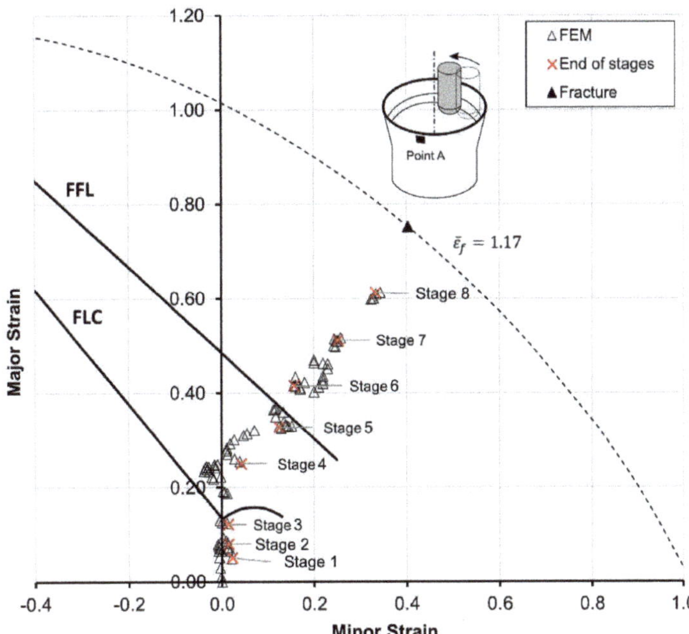

Figure 10. Finite element predicted in-plane strains of point A during the eight forming stages of incremental tube expansion.

The comparison of the results obtained for incremental tube expansion against those obtained for conventional tube expansion with a rigid tapered conical punch [9–12] allowed identifying two main differences regarding material flow and cracking. First, incremental tube expansion is performed under biaxial stretching conditions, whereas conventional tube expansion subjects the material to near pure tension. Second, both processes fail by tensile stresses (opening mode I), but while fracture in incremental tube expansion is not preceded by necking, that is not the case in conventional tube expansion, in which fracture is preceded by localized necking.

Even though all the experimental and theoretical results presented in this section are consistent and compatible, there is a fundamental problem arising from the fact that in-plane strains of incremental tube expansion are far greater than the threshold admissible values given by the FFL. Because the FFL is a material property, whose values cannot be surpassed, Cristino et al. [16] put forward the possibility of the FFL having an upward curvature in the first quadrant of principal strain space to accommodate the values in excess (i.e., to accommodate the in-plane strains located above the straight line falling from left to right), but they did not provide evidence for this type of tube material.

In connection to this, it is worth noticing that recent published works in incremental sheet forming also report the existence of strain paths that go beyond the FFL determined by means of conventional sheet forming tests with proportional strain paths [22].

The following section focuses on this problem and aims at providing an explanation for the reason why the critical strains of incremental tube expansion at fracture are far superior to those of conventional tube expansion that is simultaneously compatible with the FFL being a material property, whose threshold values cannot be surpassed by any type of loading. The explanation will make use of the effective strain vs. stress triaxiality space instead of the principal strain space.

3.2. FFL and Stress Triaxiality under Non-Proportional Paths

Damage accumulation associated to growth and coalescence of voids subjected to tensile normal stresses (Mode I) accounts for the dilatational effects related to stress triaxiality $\eta = \sigma_m / \overline{\sigma}$, in the form of a weighted integral form of the effective plastic strain [23–25].

$$D = \int_0^{\overline{\varepsilon}} \frac{\sigma_m}{\overline{\sigma}} d\overline{\varepsilon} \tag{9}$$

The critical damage D_{crit} at the onset of fracture (FFL) corresponds to the maximum admissible accumulated value of effective strain $\overline{\varepsilon} = \overline{\varepsilon}_f$ for a given strain path.

The accumulation of damage D in principal strain space often distinguishes between two different types of strain paths: (i) linear, proportional strain paths (Figure 11a) and (ii) non-proportional strain paths, which are often discretized through a series of piecewise linear strain paths for calculation purposes (Figure 11b).

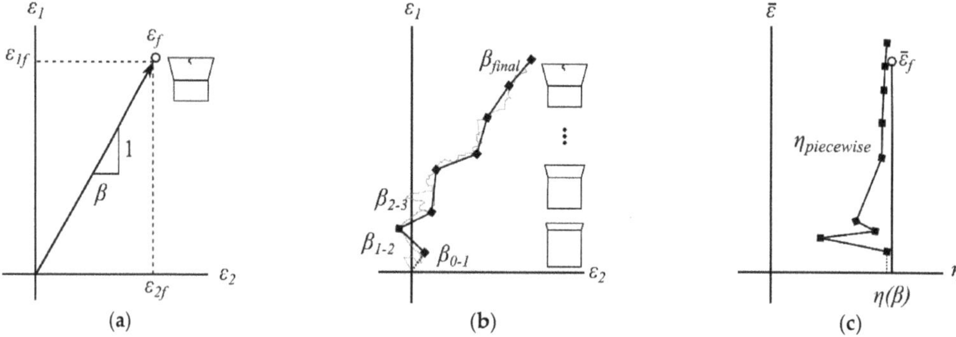

Figure 11. (a) Linear, proportional strain path in principal strain space, (b) non-proportional strain path discretized through a series of piecewise linear strain paths in principal strain space, (c) representation of the strain paths (a,b) in the effective strain vs. stress triaxiality space.

As shown in Figure 11c, the application of Equations (3) and (4) to linear, proportional strain paths, characterized by a constant slope $\beta = d\varepsilon_2 / d\varepsilon_1$ (Figure 11a), gives rise to vertical lines $\eta = \eta_p$ in the effective strain vs. stress triaxiality space. In contrast, the application of Equations (3) and (4) to non-proportional, piecewise linear strain paths with different slopes β_i, gives rise to piecewise linear evolutions $\bar{\varepsilon} = f(\eta_i)$ (hereafter referred to as $\eta_{piecewise}$ based evolutions) in the effective strain vs. stress triaxiality space.

The experimental strain paths disclosed in Figure 8 allow concluding that tube expansion by a rigid tapered conical punch subject the material to linear, proportional (or near proportional) strain paths, whereas incremental tube expansion subjects the material to non-proportional strain paths. The picture inserts of Figure 11a,b are drawn in accordance with this conclusion.

However, the strain paths determined by CGA using the automatic measurement system ARGUS® must be seen as static results obtained at the end of the incremental tube expansion process (Figure 8), or at the end of each intermediate forming stage (Figure 10). Full characterization of the non-proportional strain paths of incremental tube forming with detailed information on the cyclic oscillations from shearing to biaxial stretching, as the single point hemispherical tool approaches, contacts and moves away from a specific location of the incrementally expanded tube surface can only be obtained through finite element modelling.

Figure 12 provides a schematic representation of a finite element computed non-proportional, cyclic path undergone by a specific tube location in the effective strain vs. stress triaxiality space. Three different evolutions $\bar{\varepsilon} = f(\eta)$ are considered as a result of the following three approaches to account for the accumulation of damage D in non-proportional, cyclic paths:

(a) The envelope stress triaxiality η_{env} based approach (Figure 12a).

$$D_{env} = \eta_{env}\bar{\varepsilon} \;\rightarrow\; \eta_{env} = \frac{1}{\bar{\varepsilon}} \int_0^{\bar{\varepsilon}} \left(\frac{\sigma_m}{\bar{\sigma}}\right)_{max} d\bar{\varepsilon} \tag{10}$$

where $(\cdot)_{max}$ stands for the peak values of the stress triaxiality ratio at each cycle (circular path) of the forming tool.

(b) The average positive stress triaxiality $\bar{\eta}_{pos}$ based approach (Figure 12b).

$$D_{pos} = \bar{\eta}_{pos}\bar{\varepsilon} \;\rightarrow\; \bar{\eta}_{pos} = \frac{1}{\bar{\varepsilon}} \int_0^{\bar{\varepsilon}} \langle\frac{\sigma_m}{\bar{\sigma}}\rangle d\bar{\varepsilon} \tag{11}$$

where $\langle\cdot\rangle$ corresponds to the Macaulay bracket to prevent accumulation of negative damage.

(c) The average stress triaxiality $\bar{\eta}$ based approach (Figure 12c), where $\overline{D} = D$ of Equation (9).

$$\overline{D} = \bar{\eta}\bar{\varepsilon} \;\rightarrow\; \bar{\eta} = \frac{1}{\bar{\varepsilon}} \int_0^{\bar{\varepsilon}} \frac{\sigma_m}{\bar{\sigma}} d\bar{\varepsilon} \tag{12}$$

where $\overline{D} = D$ of Equation (9).

The three evolutions $\bar{\varepsilon} = f(\eta)$ resulting from these approaches are identical in case of linear, proportional strain paths because in such loading conditions, $D = D_{env} = D_{pos} = \overline{D}$.

Figure 13 shows the finite element non-proportional, cyclic path of incremental tube forming experienced by point A of Figure 9 and the three different $\bar{\varepsilon} = f(\eta)$ evolutions that result from the integral forms of stress triaxiality η_{env}, $\bar{\eta}_{pos}$ and $\bar{\eta}$ given by Equations (10)–(12). The linear piecewise $\eta_{piecewise}$ based evolution resulting from the experimental in-plane strains obtained by ARGUS® and by the linear, proportional, equal biaxial stretching η_p based evolution are included for comparison purposes.

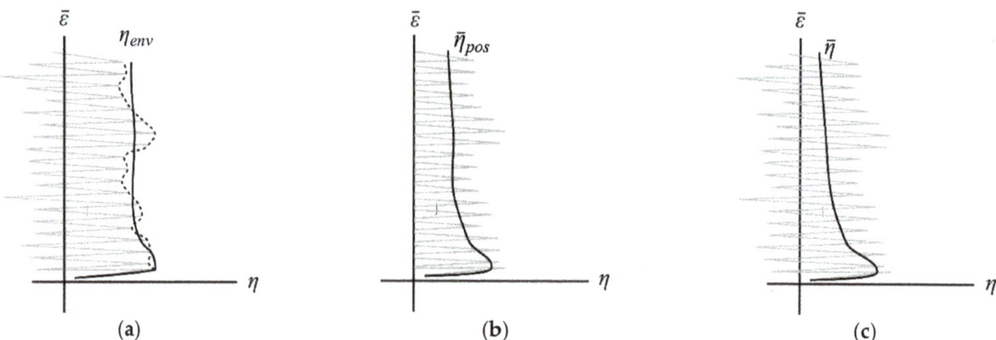

Figure 12. Schematic representation of the non-proportional, cyclic path of incremental tube expansion experienced by an arbitrary tube location with a plot of the $\bar{\varepsilon} = f(\eta)$ evolutions based on the three different integral forms of stress triaxiality: (a) envelope stress triaxiality η_{env}, (b) average positive stress triaxiality $\bar{\eta}_{pos}$ and (c) average stress triaxiality $\bar{\eta}$.

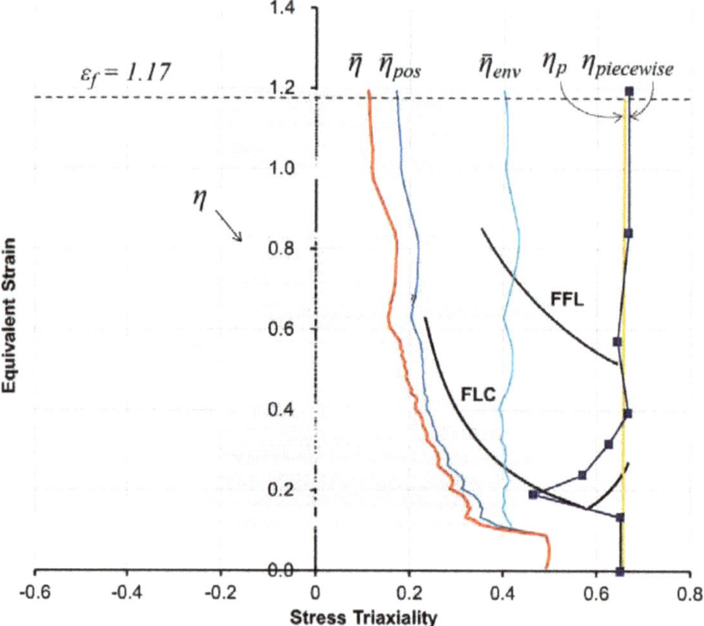

Figure 13. Finite element computed non-proportional, cyclic path of point A (Figure 9) with several $\bar{\varepsilon} = f(\eta)$ evolutions obtained from different assumptions and integral forms of stress-triaxiality.

As seen, the $\bar{\varepsilon} = f(\eta)$ reference evolution based on a linear, proportional, equal biaxial stress triaxiality ratio η_p consists of a vertical line $\eta_p = 0.66$ that extends up to an effective strain value at fracture $\bar{\varepsilon}_f = 1.17$ located far above the FFL. The other $\bar{\varepsilon} = f(\eta)$ reference evolution based on a linear piecewise $\eta_{piecewise}$ approximation of the experimental in-plane strains measured by ARGUS® is not very much different from that based on η_p. Major differences between the two evolutions are found in the forming stages 2 to 5 due to a shift in the linear piecewise $\eta_{piecewise}$ based evolution towards plane strain.

Still, the onset of fracture at $\bar{\varepsilon}_f = 1.17$ is nearly identical to that of the η_p based evolution and, therefore, far above the FFL. In fact, because the linear piecewise $\eta_{piecewise}$ based evolution is built upon a direct transformation of the experimental strain paths from principal strain space to the effective strain vs. stress triaxiality space, it is understandable that the surpass of the FFL must occur in both spaces.

More important to our discussion are the $\bar{\varepsilon} = f(\eta)$ evolutions obtained for the integral forms of stress triaxiality given by η_{env}, $\bar{\eta}_{pos}$ and $\bar{\eta}$ (refer to Equations (10)–(12)). As can be seen, the three evolutions reach the effective strain at fracture ($\bar{\varepsilon}_f = 1.17$) very far from the FFL. In particular, the evolution of η_{env} cuts the FFL at stress triaxiality values around 0.4, suggesting that the fracture should occur much earlier than it does. The other two ($\bar{\eta}_{pos}$ and $\bar{\eta}$) reach the fracture for values of stress triaxiality below 0.2 (i.e., in-between pure tension and pure shear) without crossing the FFL and in good agreement with a possible extrapolation of the FFL to the left side. The difference between the $\bar{\eta}_{pos}$ and $\bar{\eta}$ based evolutions is not relevant for incremental tube expansion and derives from discharging, or accounting for, the accumulation of negative damage. Although discharging negative damage is commonly executed in cold forming, there are studies recently published pointing to cut-off values of stress triaxiality up to -0.6 for the cold forming of aluminum alloys under quasi-static loading [26]. According to this and taking into account that the instantaneous stress triaxiality in the incremental tube forming oscillates between -0.6 to 0.6 (see Figure 13), the use of the average stress triaxiality $\bar{\eta}$ takes on a greater physical sense.

Taking the integral form of stress triaxiality $\bar{\eta}$ (i.e., the average stress triaxiality given by Equation (12)) into consideration, it is now important to check if the compatibility between the FFL and the above-mentioned reason for the critical in-plane strains of incremental tube expansion at fracture being far greater than those of conventional tube expansion also applies to the latter. For this purpose, we computed the $\bar{\varepsilon} = f(\eta)$ evolution for conventional tube expansion directly from the average stress triaxiality $\bar{\eta}$ and plotted the results in Figure 14. The instantaneous stress triaxiality (η) in the conventional tube expansion is also shown.

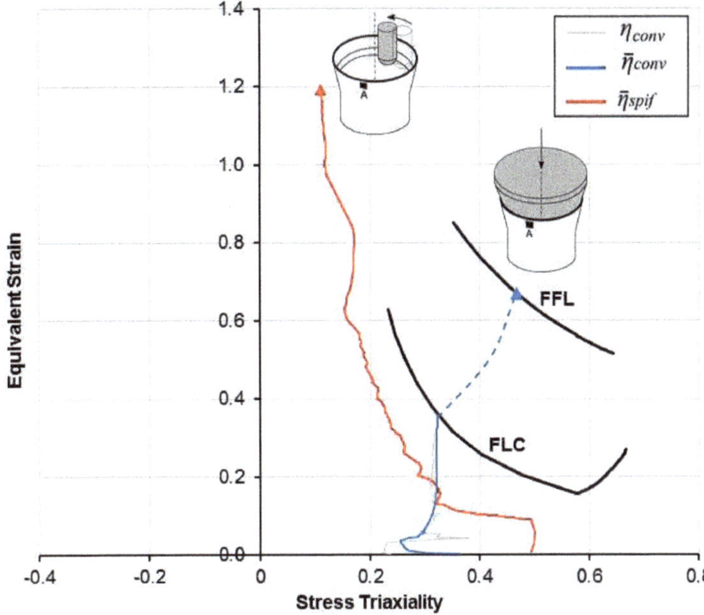

Figure 14. Finite element computed evolutions of the loading paths experienced by a point A located 1.5 mm away from the upper tube end in incremental and conventional tube expansion processes. Note: the red and blue triangular markers correspond to the experimentally determined "gauge length" strains at fracture.

Two interesting results can be drawn. On the one hand, the level of average stress triaxiality in the conventional process at fracture ($\overline{\eta}_f \approx 0.47$) is very well above the one obtained in the incremental process ($\overline{\eta}_f \approx 0.11$). As suggested by Martinez-Donaire et al. [18], this difference, also observed in other incremental forming processes [27], results in a greater resistance to accumulate damage in the incremental process than in the conventional one, requiring higher levels of strain to trigger the ductile fracture. On the other hand, results also show a near coincidence of the instantaneous and the average stress triaxiality-based evolutions ($\eta \cong \overline{\eta}$) in the conventional process. This also makes sense and is compatible with the widely popular application of McClintock's ductile damage criterion [23] to determine the onset of cracking by tension in conventional tube forming processes [12]. These results confirm the validity of the overall approach for both non-proportional, cyclic paths of incremental tube expansion and near proportional paths of conventional tube expansion.

4. Conclusions

The critical in-plane strains and effective strain at fracture of incremental tube expansion are greater than those of conventional tube expansion by rigid tapered conical punches. However, the conclusion that the gains in formability are due to the fact that strain paths go beyond the fracture forming limit (FFL) line of the tube material is erroneous because it does not account for the non-proportional, cyclic nature of the strain paths and because it ignores the FFL being a material property that is independent of any type of applied loading.

Finite element modelling of incremental tube expansion considering material strain hardening and non-proportional, cyclic paths resulting from the real tool trajectory combined with the utilization of appropriate integral forms of stress triaxiality allows understanding that the gains in formability result from a shift of the loading paths towards the left side in the effective strain vs. stress triaxiality space. Moreover, the results also show that this shift of material flow is of paramount importance to ensure compatibility between the critical strains at fracture and the threshold admissible values of the material FFL.

The necessity of using effective strain vs. stress triaxiality evolutions based on average stress triaxiality to ensure compatibility with the FFL in incremental tube expansion is understandable because the individual locations of the plastically deformed tube surface oscillate cyclically from shearing to biaxial stretching, as the single point hemispherical tool approaches, contacts and moves away from these locations during its trajectory.

The fact that average stress triaxiality is not required to handle the formability of conventional tube expansion is compatible with the match between stress triaxiality and the integral forms of stress triaxiality (e.g., average stress triaxiality) when material is subject to near-proportional loading paths. This last conclusion is no less important than the previous ones because it justifies the successful utilization of McClintock's fracture criterion to analyze the onset of fracture by tension in conventional tube forming processes over the past few decades.

Author Contributions: Conceptualization, G.C., C.V. and P.A.F.M.; methodology, M.B.S. and G.C.; software, C.S.; validation, C.S. and G.C.; formal analysis, G.C. and M.B.S.; investigation, C.S., M.B.S. and G.C.; resources, C.V., M.B.S. and P.A.F.M.; writing—original draft preparation, C.S. and G.C.; supervision, G C. and C.V.; project administration, G.C. and C.V.; funding election and management, G.C. All authors have read and agreed to the published version of the manuscript.

Funding: The authors would like to express their gratitude for the funding received through the Grant number US-1263138 US/JUNTA/FEDER_UE within "Proyectos I+D+i FEDER Andalucía 2014-2020" and the Fundação para a Ciência e da Tecnologia of Portugal, through IDMEC under LAETA, project UIDB/50022/2020.

Institutional Review Board Statement: Not applicable.

Informed Consent Statement: Not applicable.

Data Availability Statement: Not applicable.

Acknowledgments: The authors would like to acknowledge the collaborations of Jaime Blanco Vital in the preliminary development of the numerical modeling and Valentino Cristino in the experimentation.

Conflicts of Interest: The authors declare no conflict of interest.

References

1. Keeler, S.P. Circular grid system–A valuable aid for evaluating sheet metal formability. In *SAE Transactions*; SAE International: Warrendale, PA, USA, 1968.
2. Goodwin, G.M. Application of strain analysis to sheet metal forming problems in the press shop. In *SAE Transactions*; SAE International: Warrendale, PA, USA, 1968.
3. Embury, J.D.; Duncan, J.L. Formability Maps. *Annu. Rev. Mater. Sci.* **1981**, *11*, 505–521. [CrossRef]
4. Centeno, G.; Martínez-Donaire, A.J.; Morales-Palma, D.; Vallellano, C.; Silva, M.B.; Martins, P.A.F. Novel experimental techniques for the determination of the forming limits at necking and fracture. *Mater. Form. Mach. Res. Dev.* **2015**, *2*, 1–24.
5. Nielsen, C.V.; Martins, P.A.F. *Metal Forming: Formability, Simulation, and Tool Design*; Elsevier Academic Press: Cambridge, MA, USA, 2021.
6. Isik, K.; Silva, M.B.; Tekkaya, A.E.; Martins, P.A.F. Formability limits by fracture in sheet metal forming. *J. Mater. Process. Technol.* **2014**, *214*, 1557–1565. [CrossRef]
7. Liu, S.D.; Meuleman, D.; Thompson, K. Analytical and experimental examination of tubular hydroforming limits. *J. Mater. Manuf.* **1998**, *107*, 287–297.
8. Chen, X.; Yu, Z.; Hou, B.; Li, S.; Lin, Z. A theoretical and experimental study on forming limit diagram for a seamed tube hydroforming. *J. Mater. Process. Technol.* **2011**, *211*, 2012–2021. [CrossRef]
9. Centeno, G.; Silva, M.B.; Alves, L.M.; Vallellano, C.; Martins, P.A.F. Towards the characterization of fracture in thin-walled tube forming. *Int. J. Mech. Sci.* **2016**, *119*, 12–22. [CrossRef]
10. Cristino, V.A.; Magrinho, J.P.; Centeno, G.; Silva, M.B.; Martins, P.A. A digital image correlation based methodology to characterize formability in tube forming. *J. Strain Anal. Eng. Des.* **2019**, *54*, 139–148. [CrossRef]
11. Magrinho, J.P.; Centeno, G.; Silva, M.B.; Vallellano, C.; Martins, P.A.F. On the formability limits of thin-walled tube inversion using different die fillet radii. *Thin-Walled Struct.* **2019**, *144*, 106328. [CrossRef]
12. Magrinho, J.P.; Silva, M.B.; Centeno, G.; Moedas, F.; Vallellano, C.; Martins, P.A.F. On the determination of forming limits in thin-walled tubes. *Int. J. Mech. Sci.* **2019**, *155*, 380–391. [CrossRef]
13. Magrinho, J.P.; Silva, M.B.; Martins, P.A.F. On the Characterization of Fracture Loci in Thin-Walled Tube Forming. In *Forming the Future, Proceedings of the 13th International Conference on the Technology of Plasticity, Pittsburgh, PA, USA, 25–30 July 2021*; Springer International Publisher: Berlin/Heidelberg, Germany, 2021.
14. Wen, T.; Yang, C.; Zhang, S.; Liu, L. Characterization of deformation behavior of thin-walled tubes during incremental forming: A study with selected examples. *Int. J. Adv. Manuf. Technol.* **2015**, *78*, 1769–1780. [CrossRef]
15. Movahedinia, H.; Mirnia, M.J.; Elyasi, M.; Baseri, H. An investigation on flaring process of thin-walled tubes using multistage single point incremental forming. *Int. J. Adv. Manuf. Technol.* **2018**, *94*, 867–880. [CrossRef]
16. Cristino, V.A.; Magrinho, J.P.; Centeno, G.; Silva, M.B.; Martins, P.A.F. Theory of single point incremental forming of tubes. *J. Mater. Process. Technol.* **2021**, *287*, 116659. [CrossRef]
17. Vujovic, V.; Shabaik, A.H. A new workability criterion for ductile metals. *J. Eng. Mater. Technol. Trans. ASME* **1986**, *108*, 245–249. [CrossRef]
18. Martínez-Donaire, A.J.; García-Lomas, F.J.; Vallellano, C. New approaches to detect the onset of localised necking in sheets under through-thickness strain gradients. *Mater. Des.* **2014**, *57*, 135–145. [CrossRef]
19. Martínez-Donaire, A.J.; Borrego, M.; Morales-Palma, D.; Centeno, G.; Vallellano, C. Analysis of the influence of stress triaxiality on formability of hole-flanging by single-stage SPIF. *Int. J. Mech. Sci.* **2019**, *151*, 76–84. [CrossRef]
20. Guglielmi, P.; Cusanno, A.; Bagudanch, I.; Centeno, G.; Ferrer, I.; Garcia-Romeu, M.L.; Palumbo, G. Experimental and numerical analysis of innovative processes for producing a resorbable cheekbone prosthesis. *J. Manuf. Process.* **2021**, *70*, 1–14. [CrossRef]
21. Centeno, G.; Martínez-Donaire, A.J.; Bagudanch, I.; Morales-Palma, D.; Garcia-Romeu, M.L.; Vallellano, C. Revisiting formability and failure of AISI304 sheets in SPIF: Experimental approach and numerical validation. *Metals* **2017**, *7*, 513. [CrossRef]
22. Mirnia, M.J.; Shamsari, M. Numerical prediction of failure in single point incremental forming using a phenomenological ductile fracture criterion. *J. Mater. Process. Technol.* **2017**, *244*, 17–43. [CrossRef]
23. McClintock, F.A. A Criterion for Ductile Fracture by the Growth of Holes. *J. Appl. Mech.* **1968**, *35*, 363–371. [CrossRef]
24. Atkins, A.G. Fracture in forming. *J. Mater. Process. Technol.* **1996**, *56*, 609–618. [CrossRef]
25. Martins, P.A.F.; Bay, N.; Tekkaya, A.E.; Atkins, A.G. Characterization of fracture loci in metal forming. *Int. J. Mech. Sci.* **2014**, *83*, 112–123. [CrossRef]

26. Brünig, M.; Gerke, S.; Schmidt, M. Damage and failure at negative stress triaxialities: Experiments, modeling and numerical simulations. *Int. J. Plast.* **2018**, *102*, 70–82. [CrossRef]
27. López-Fernández, J.A.; Centeno, G.; Martínez-Donaire, A.J.; Morales-Palma, D.; Vallellano, C. Stretch-flanging of AA2024-T3 sheet by single-stage SPIF. *Thin-Walled Struct.* **2021**, *160*, 107338. [CrossRef]

Article

A New and Direct R-Value Measurement Method of Sheet Metal Based on Multi-Camera DIC System

Siyuan Fang [1], Xiaowan Zheng [1,2], Gang Zheng [1], Boyang Zhang [1], Bicheng Guo [1] and Lianxiang Yang [1,*]

[1] Department of Mechanical Engineering, School of Engineering and Computer Science, Oakland University, Rochester, MI 48309, USA; siyuanfang@oakland.edu (S.F.); xiaowanzheng@oakland.edu (X.Z.); gangzheng@oakland.edu (G.Z.); boyangzhang@oakland.edu (B.Z.); bichengguo@oakland.edu (B.G.)

[2] College of Electrical and Information Engineering, Zhengzhou University of Light Industry, Zhengzhou 450002, China

* Correspondence: yang2@oakland.edu

Abstract: More and more attention has been given in the field of mechanical engineering to a material's R-value, a parameter that characterizes the ability of sheet metal to resist thickness strain. Conventional methods used to determine R-value are based on experiments and an assumption of constant volume. Because the thickness strain cannot be directly measured, the R-value is currently determined using experimentally measured strains in the width, and loading directions in combination with the constant volume assumption, to determine the thickness strain indirectly. This paper provides an alternative method for determining the R-value without any assumptions. This method is based on the use of a multi-camera DIC system to measure strains in three directions simultaneously. Two sets of stereo-vision DIC measurement systems, each comprised of two GigE cameras, are placed on the front and back sides of the sample. Use of the double-sided calibration strategy unifies the world coordinate system of the front and back DIC measurement systems to one coordinate system, allowing for the measurement of thickness strain and direct calculation of R-value. The Random Sample Consensus (RANSAC) algorithm is used to eliminate noise in the thickness strain data, resulting in a more accurate R-value measurement.

Keywords: R-value; thickness strain; digital image correlation; multi-camera DIC; non-destructive testing

Citation: Fang, S.; Zheng, X.; Zheng, G.; Zhang, B.; Guo, B.; Yang, L. A New and Direct R-Value Measurement Method of Sheet Metal Based on Multi-Camera DIC System. *Metals* **2021**, *11*, 1401. https://doi.org/10.3390/met11091401

Academic Editors: Gabriel Centeno and Ricardo Jose Alves de Sousa

Received: 10 July 2021
Accepted: 2 September 2021
Published: 4 September 2021

Publisher's Note: MDPI stays neutral with regard to jurisdictional claims in published maps and institutional affiliations.

1. Introduction

The plastic strain ratio r is a parameter that indicates the ability of a sheet metal to resist thinning or thickening when subjected to either tensile or compressive forces in the plane of the sheet [1,2]. It has been used as an important parameter to evaluate the formability of automotive sheet metal. It is typically advantageous for the material to experience a minimal reduction in the area when subjected to a force; this means that it is a good drawing material with a high R-value. Therefore, an accurate measurement of R-value is of great significance to the study of the tensile and compressive properties of materials. Stickels [3] predicted the plastic-strain ratio of low-carbon steel sheets using the Young's Modulus, while Ghosh [4] calculated the R-value of Al-Mg-Si alloy sheets based on crystal plasticity models. However, since the crystal structure of polycrystalline materials is often difficult to model accurately, the accuracy of the current theoretical estimates is not high [5,6]. Some scholars [7–9] employed laser-ultrasound resonance spectroscopy to measure the texture, thickness, and plastic strain ratio on-line. Although the laser-ultrasonic method has the advantages of being fast and non-destructive, the detection accuracy is greatly affected by the vibration of the tested strip and changes in the environment temperature. In some traditional R-value measurement methods [10,11], the thickness strain is calculated using length strain and width strain. The stated measurement methods are based on the assumption of constant volume, which states that the volume of the material will not change during the deformation process. However, some materials do not always follow the constant volume assumption, especially inside the necking band. This makes the exact determination of the R-value challenging [12]. Because of this, development of an experimental method for direct measurement of R-value becomes meaningful.

Digital Image Correlation (DIC), a non-contact measurement method, has achieved great success in the field of optical measurement and is still developing rapidly [13–17]. A first attempt to experimentally measure R-value using DIC was reported by Xie in 2017 [18]. Instead of observing the front surface of the sheet metal, this method observed the sheet metal at a 45-degree angle. As a result, both the front surface and the depth side of the sheet metal can be observed at the same time by two DIC cameras, and the three strains ε_1, ε_2 and ε_3 can be measured simultaneously. A shortcoming of this method is that the thickness strain can only be measured on the edge. Additionally, the edge needs to be cut, which will introduce a residual strain and ultimately affect the measurement results.

Thickness strain measurement at any point on the sheet metal surface using DIC was first reported by our lab at Oakland University together with Dantec-Dynamics GmbH, Germany [19]. This method was based on using multiple multi-camera DIC systems to measure strains in three directions simultaneously. Two sets of stereo vision DIC measurement systems, each consisting of two GigE cameras, were used in this method. One stereo vision DIC system was placed on the front side of the test sample, while the other system was placed on the back side of the sample. The use of the double-sided calibration strategy unified the world coordinate system of the front and back DIC measurement systems to one coordinate system, thus enabling the measurement of the thickness strain. This paper describes the use of this technology to measure the R-value directly without any assumptions. A tensile test of DP980 was conducted to directly measure the R-value. The theory of multi-camera DIC and the experimental setup will briefly be explained. Section 2 of this paper explains the DIC related theories, the two-sided calibration strategy, and the algorithms to smooth the thickness strain, while Section 3 presents the experimental setup and experimental process. The two R-value calculation methods will be presented, as well as the experimental results of DP980 and a comparison between the R-value determined through experimental results determined using the constant volume assumption. The potential and limitation of the method, as well as conclusions, will be discussed and presented in Section 4.

2. Fundamental of DIC

2.1. Basic Theory

The stereo vision system shown in Figure 1, or a two-camera system with different perspectives, plays an important role in the DIC measurement system. A space point p is recorded simultaneously by two cameras on the image plane with a different perspective. The points $p1$ and $p2$ on the image plane of the two cameras are then matched using the DIC matching algorithm, and the 3D coordinates of point p can be recovered based on the geometry of the stereo-vision system.

The core idea of the DIC method is to measure displacement and strain information on an object's surface by matching and tracking feature points in the natural texture or artificial speckle pattern before and after deformation. The DIC matching process obtains the coordinate on the target image after deformation corresponding to a known coordinate position on the reference image taken before deformation. It is usually necessary to divide the image into multiple grids, or subsets, and the deformation of the subset is used to represent the deformation of the local area of the object. Figure 2 shows this principle. On the reference image, a rectangle with a size of $(2M + 1)$ pixel $\times$ $(2M + 1)$ pixel and point $p(x_0, y_0)$ at the center is taken as the reference subset. A correlation function is then used to find a subset on the target image with a similar distribution to that of the reference subset. The subset with the largest correlation coefficient is considered the deformed subset. The sum squared difference correlation criterion (SSD) is commonly used to evaluate the similarity between subsets [20]. Assuming that the intensity distribution of the two facets is represented as $F(x, y)$ and $G(x', y')$, the SSD function has the following form:

$$C_{SSD} = \sum F(x,y) \times G(x',y') \tag{1}$$

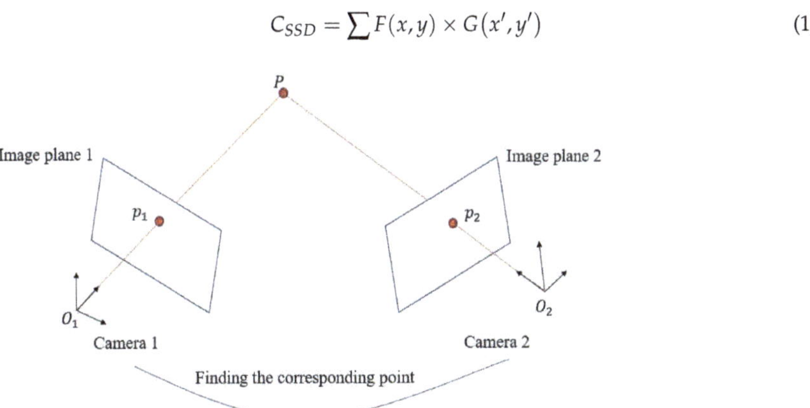

Figure 1. Schematic of stereo-vision system.

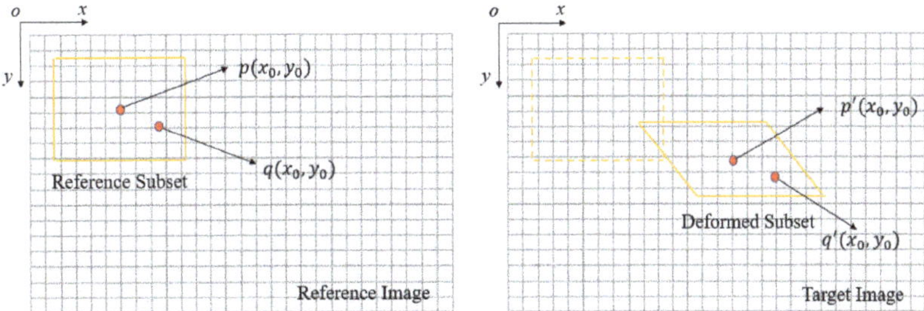

Figure 2. Basic principle of digital image correlation.

2.2. Double-Sided Calibration

Camera calibration plays an important role in DIC measurement, as its quality directly affects measurement accuracy. The principle of camera calibration is to establish the relationship between the position of the camera image pixel and the scene point position. According to the camera imaging model, the relationship between the image coordinate system and the world coordinate system is as follows [21]:

$$s\vec{m} = A\,[r\ t]\,\vec{M} \tag{2}$$

where $\overline{m} = [u,v,1]^T$ is the image homogeneous coordinate system, $\overline{M} = [X,Y,Z,1]^T$ is the world homogeneous coordinate system, s is the scale factor, A is the camera internal parameter matrix, and $[r\ t]$ is the camera external parameter matrix. Both the internal and external parameters of the camera can be calculated by Zhang's calibration method [21].

The double-sided calibration strategy uses a double-sided calibration plate to calibrate the front and back dual-camera DIC systems simultaneously. The schematic of the double-sided calibration strategy is shown in Figure 3. Figure 3 shows two typical two-camera stereo vision subsystems, and these two subsystems are linked through calibration. Figure 4 shows the design of the calibration plate. The role of this calibration is to connect these two subsystems into one single global system. The world coordinate system of the front camera is taken as the reference global coordinate, and the back camera coordinate system is transformed to the reference global system. This transformation can be expressed as follows:

$$\vec{M} = \vec{m}_{world}^{\,front} = \vec{m}_{world}^{\,back} \cdot T_\omega \quad T_\omega = \begin{bmatrix} -1 & 0 & 0 & 0 \\ 0 & 1 & 0 & 0 \\ 0 & 0 & -1 & d \\ 0 & 0 & 0 & 1 \end{bmatrix} \tag{3}$$

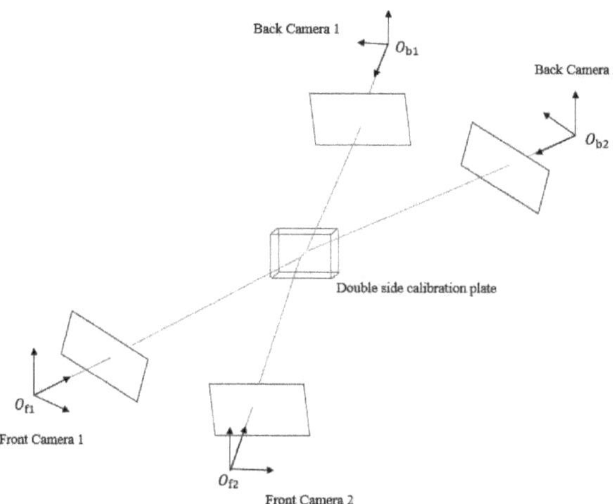

Figure 3. Schematic of double-sided calibration strategy.

Figure 4. Double side calibration plate (front and back side).

In this equation, d is the thickness of the double-sided calibration plate.

2.3. Optimized Thickness Strain Measurement

The measurement of the thickness strain is the core part of calculating the R-value. Li built a dual-camera DIC system on the front and back of the test piece to realize the non-destructive measurement of a sample's thickness strain [19]. However, as Li wrote in [19], the strain measurement results contain noise, so it is necessary to remove the noise from the strain data. Since the thickness of the test sample is usually a few millimeters, the noise is amplified during the thickness strain measurement. Figure 5 shows the DP980 thickness strain history, where the x-axis represents the number of photos taken and the yellow circle indicates the noise data. Two smoothing algorithms, namely the Least squares algorithm and Random sample consensus (RANSAC) algorithm, were used to fit the curve and reduce noise. A comparison of the results using these two algorithms follows.

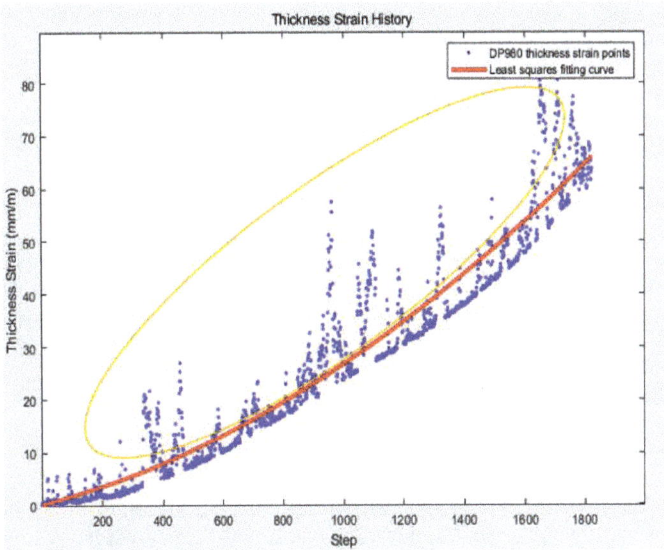

Figure 5. DP980 thickness strain fitted by least squares algorithm.

2.3.1. Least Squares Algorithm

The least squares method is the most common method used to solve curve fitting problems. The basic idea of the method is to minimize the sum of squares of the error to find the best function match for the data. In general, we can model the expected value of y as an nth degree polynomial, yielding the general polynomial regression model:

$$y = f(x, \alpha) = \alpha_0 + \alpha_1 x + \alpha_2 x^2 + \alpha_3 x^3 + \cdots \alpha_n x^n \tag{4}$$

where $\alpha = [\alpha_1, \alpha_2, \cdots, \alpha_n]$ is a parameter to be determined.

The objective function $L(y, f(x, \alpha))$ is minimized to find the optimal estimated value of the parameter α in the function $f(x, \alpha)$ for m given sets of data (x_i, y_i) $(i = 1, 2, \cdots, m)$.

$$L(y, f(x, \alpha)) = \sum_{i=1}^{m} [y_i - f(x_i, y_i)]^2 \tag{5}$$

The result using least squares fitting is shown in Figure 5 as a red curve. The red curve represents the results of quadratic polynomial fitting. Observation of Figure 5 indicates that the fitted curve shifts relative to the expected data. Because of this, the algorithm's fitting results become inaccurate if the data contains a large amount of error or noise. This is the limitation of the least squares algorithm.

2.3.2. Random Sample Consensus (RANSAC) Algorithm

The RANSAC algorithm based on the iterative method was proposed by Fischler [22] to solve the problem of inaccurate solutions when using the least squares model on sample data with a large proportion of outliers. Song used the improved RANSAC algorithm to eliminate errors caused by high-temperature heat wave disturbances in DIC measurement [23]. This algorithm is able to classify data points as outliers or inliers and fit the mathematical model through inliers while ignoring outliers. The basic algorithm is summarized as follows:

1. Randomly select several points in the thickness-strain data to calculate the polynomial thickness-strain history model.
2. Set a preset tolerance value ε and calculate the number of data points that match the mathematical model in Step 1.
3. If the proportion of data points that meet the mathematical model in Step 2 exceeds the preset threshold τ, recalculate the thickness-strain history model to use these inliers and terminate the algorithm.
4. Otherwise, repeat Steps 1 to 3 K times.

The number of iterations K should be large enough to ensure that the probability p of at least one set of random samples not including outliers is greater than 0.99. Assuming that the probability that the data is selected as an inlier is a, then $b = 1 - a$ is the probability of observing an outlier. The required number of iterations of the minimum number of points, denoted k, is calculated using the following function:

$$1 - p = (1 - a^k)^K \tag{6}$$

$$K = \frac{\log(1 - p)}{\log(1 - (1 - b)^k)} \tag{7}$$

Figure 6 shows that the RANSAC algorithm classifies the thickness strain data into inlier points and outlier points. The blue point set represent the inlier points, and the red circles represent the outlier point set. Using this method, the outlier points, or noisy data, can be ignored during fitting. The result of the RANSAC algorithm fitting is shown in Figure 7. It can be seen that the red fitting curve is fitted with inlier points as the data set. Compared with the least squares algorithm, it is hardly affected by noise in the data. This indicates that the RANSAC algorithm is effective and has important significance for the subsequent R-value calculation.

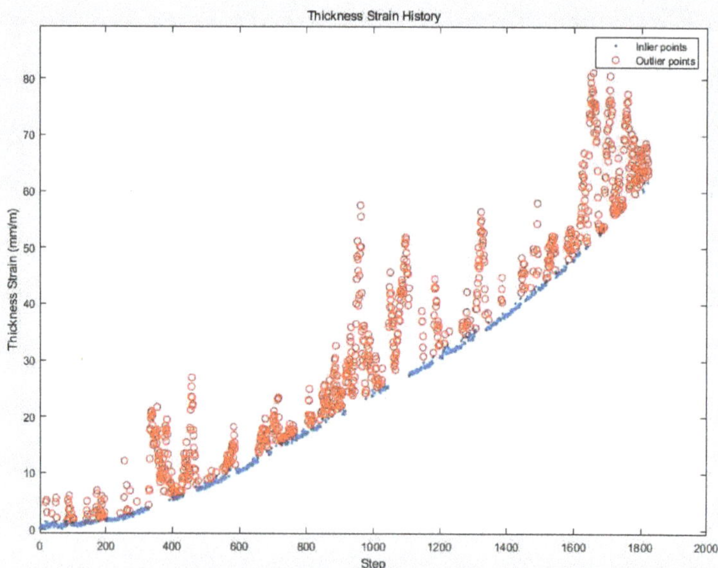

Figure 6. Thickness strain data is classified inlier and outlier.

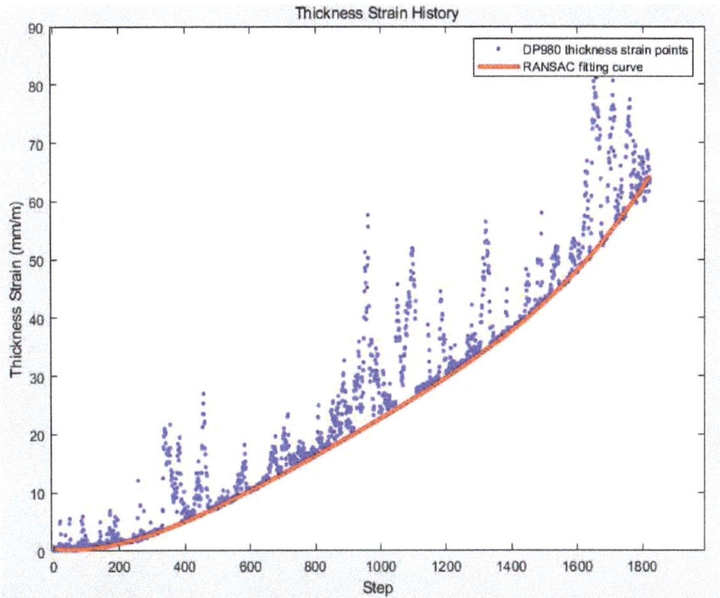

Figure 7. Fitting curve by RANSAC algorithm.

3. DP980 R-Value Determination

In this section, the R-value calculation method is reviewed first, followed by a discussion of the experimental setup and process. The composite material DP980 was used as the experimental sample. Finally, the experimental results are analyzed and discussed.

3.1. R-Value Calculation

The equation to calculate R-value is shown as Equation (8).

$$R = \frac{\varepsilon_2}{\varepsilon_3} \tag{8}$$

where $\varepsilon_2, \varepsilon_3$ represent the true width strain and true thickness strain, respectively.

For sheet metals, the R-value is usually determined from three different directions of in-plane loading ($0°, 45°, 90°$ in the rolling direction) and the normal R-value is calculated using the following equation:

$$R = \frac{1}{4}(R_0 + 2R_{45} + R_{90}) \tag{9}$$

3.2. Experiment System

The experimental system is shown in Figures 8 and 9. Four GigE cameras (maximum resolution 2752×2200) were used to form two independent DIC binocular measurement systems. One DIC system was placed in front of the test sample, while the other DIC system was placed behind the test sample. Two pairs of 50 mm fixed focus lenses were used to achieve a 20 mm × 20 mm field of view. The spatial resolution was 0.08 mm/pixel and the acquisition rate was 10 Hz. Two powerful LED lights provided sufficient illumination for the front and back DIC measurement systems. A 50 KN MTS tensile test machine was used to do the tensile test, and a 1 mm thick dog bone shape sample made from DP980 sheet metal was prepared for the test. The tensile speed was set at 3 mm/min. The slow tensile rate used for the test enabled more images to be taken before the sample cracked, thus providing more data to analyze. Figure 10 is a schematic diagram of this sample. Current limitations with this system include the cost of the cameras used for the DIC systems and the large amount of space required to accommodate the front and rear camera groups.

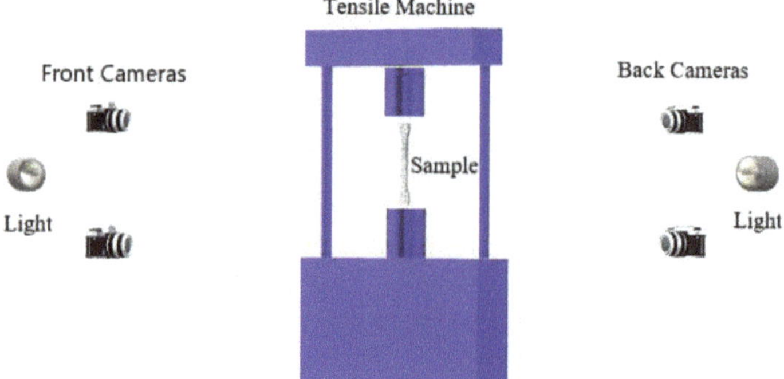

Figure 8. Schematic diagram of the experimental system.

Figure 9. Experimental set-up of the R-value measurement.

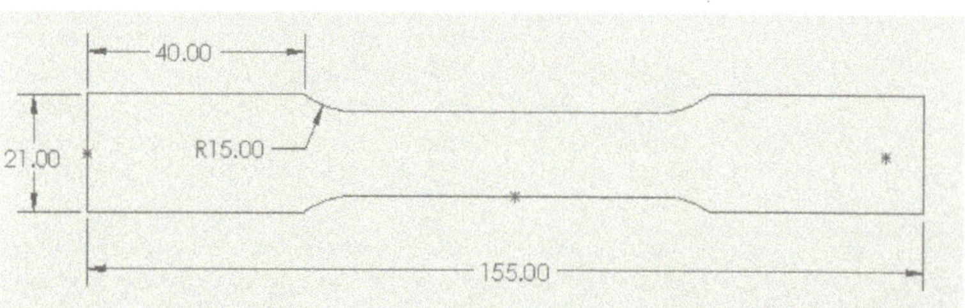

Figure 10. The schematic diagram of DP 980.

3.3. Experimental Results Analysis and Discussion

The 3D contour and strain distribution on the surfaces of the DP980 was obtained through DIC measurement. Figure 11 illustrates the simultaneous measurement results of length strain (ε_1) map, width strain (ε_2) map and thickness strain (ε_3) map before fracture in the tensile test. The necking area, which is indicative of a strain concentration, is clearly visible.

R-value measurements were taken using two different methods, and these results were compared and analyzed. The first method is the traditional measurement, or indirect measurement, method based on the constant volume assumption. According to the assumption of constant volume, $\varepsilon_1 + \varepsilon_2 + \varepsilon_3 = 0$; thus, the thickness strain can be derived from measurements of the length and width strain:

$$\varepsilon_3 = -(\varepsilon_1 + \varepsilon_2) \tag{10}$$

Based on the definition of R-value:

$$R = \frac{\varepsilon_2}{-(\varepsilon_1 + \varepsilon_2)} \tag{11}$$

The other method used is the direct measurement method, or the direct thickness strain measurement method proposed in this paper. The calculation method is shown in Equation (8). Data in both the non-necking and necking area was selected for analysis to better show the difference between the two measurement methods.

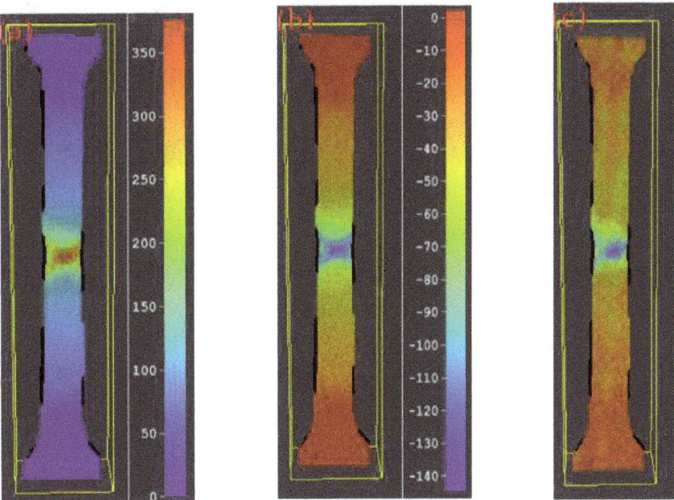

Figure 11. Full field strain distribution of sample surface measured by DIC system: (**a**) distribution of length strain; (**b**) distribution of width strain; (**c**) distribution of thickness strain.

3.3.1. Non-Necking Area

A square area of 8×8 pixels, as shown in Figure 12a, was selected instead of a single point in the non-necking area to reduce the influence of noise on the data. Each pixel represents 0.2 mm. The calculation results for the traditional and proposed methods are shown in Figure 13. According to the definition of R-value, the R-value for DP980 is equal to the slope of these two curves. It can be seen from the figure that the relationship between thickness strain and width strain is approximately linear, and the amount of noise in the data is low. Therefore, the least squares principle can be used to calculate the slope of the two curves. The directly measured R-value is 0.8546, while the indirectly measured R-value is 0.8656. Based on previous research of the properties of DP980 material, its

R-value generally ranges between 0.7–1.0, depending on the material's composition [24]. The results measured by both methods can be considered reasonable because they both fall within this range, and the difference in the results between the two methods is very small. Therefore, both methods are acceptable.

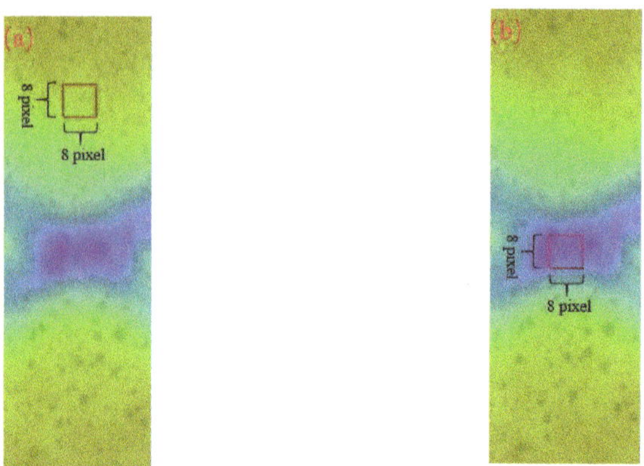

Figure 12. Calculation area of R-value: (**a**) Non-necking area; (**b**) Necking area.

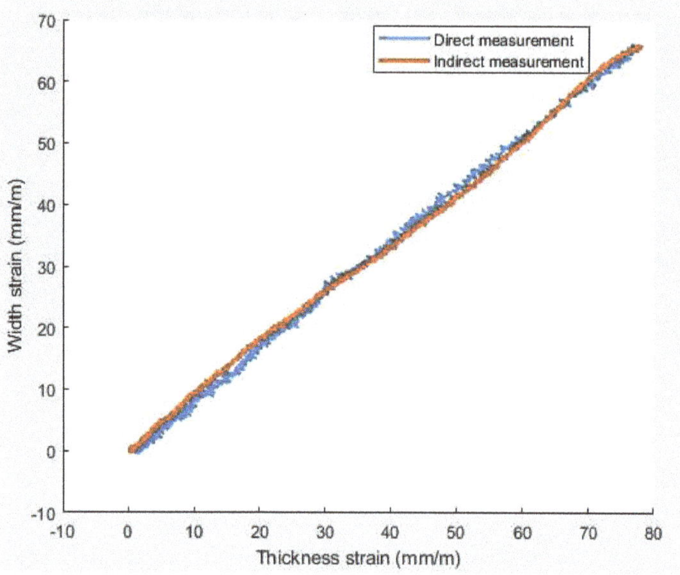

Figure 13. The R-value curve by two different measurement methods in the non-necking area.

3.3.2. Necking Area

When necking occurs during a tensile test, a large amount of strain is disproportionately distributed in a central location of the material. As the deformation continues, the strain in the necking area continues to increase until the material ruptures. In the necking stage, the cross-sectional area of the material decreases drastically, while the strain in the thickness and width directions increases sharply. Figure 14 displays an example of the sharp strain increase that can occur during the necking stage.

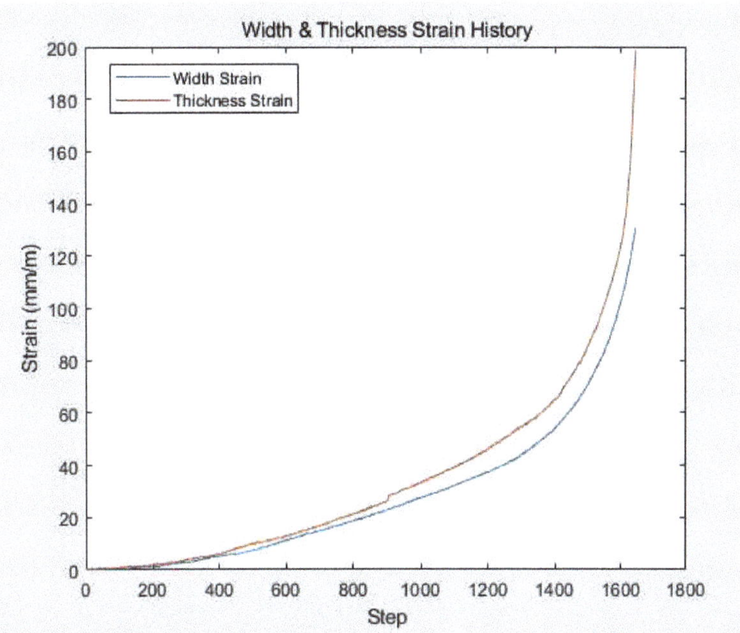

Figure 14. Width and Thickness strain history.

An 8 × 8 pixel square area was selected in the necking area to obtain strain data, as shown in Figure 12b. Figure 15 illustrates the R-value measured by the two methods in the necking area. In the linear elastic deformation stage, the thickness strain is proportional to the width strain. The R-value at this time is a constant, and the results measured by the two methods are 0.8471 and 0.8526, respectively; these results are similar to those measured in the non-necked area. This shows that the material is uniformly distributed, stable in nature, and isotropic. When the width strain exceeds 60 mm/m, the strain value increases rapidly, and the material deformation enters the necking stage. It is worth noting that the direct method measurement curve no longer increases linearly; the growth rate reduces, and the R-value correspondingly becomes smaller. This shows that the ability of the material to resist strain in the thickness direction is reduced. Compared to the curve for the direct measurement method, the curve for the indirect measurement method remains linear in the early necking stage, and the R-value does not decrease significantly until the later necking stages. This result further confirms some researcher opinions that the material does not necessarily follow the constant volume assumption in the necking stage [12]. Therefore, the R-value measured by the direct measurement method is more reliable and accurate in the necking stage. It should be noted that some materials do follow the assumption of constant volume in the necking stage. The R-values for DP980 in the three rolling directions are displayed in Table 1. The R-values in the three rolling directions are very similar, indicating that the material has similar resistance to thickness deformation in different rolling directions.

3.4. Additional Verification Tests

Two additional materials were used for uniaxial tensile tests to verify the repeatability of the method proposed in this paper: aluminum alloy 6061 and a polymer material.

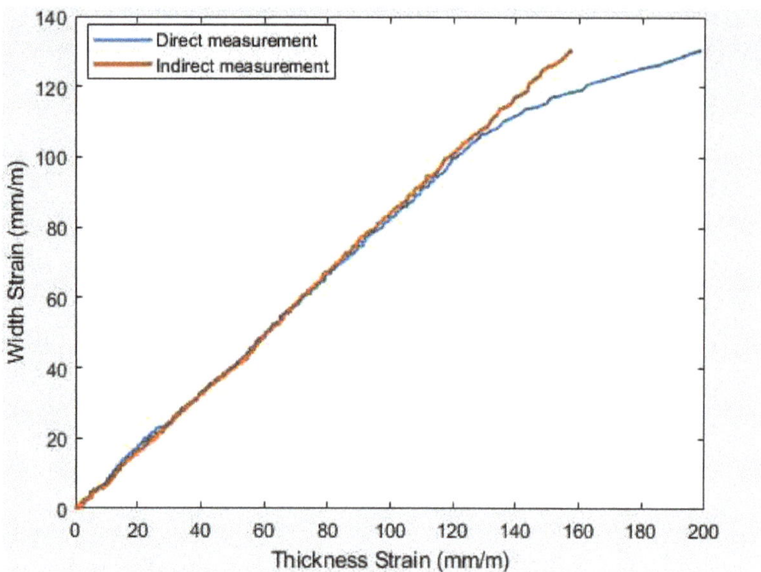

Figure 15. The R-value curve by two different measurement methods in the necking area.

Table 1. R-value of DP980 in different rolling direction.

Rolling Direction	R-Value
0°	0.8656
45°	0.8734
90°	0.8853
Sheet Metal	0.8749

3.4.1. Aluminum Alloy 6061

A 3 mm thick dog-bone shaped aluminum alloy 6061 sample was utilized for a tensile test, and the test conditions were consistent with those used for DP980. The non-necking and necking areas were used to analyze the R-value of aluminum alloy 6061. Figures 16 and 17 show the R-value curves of the non-necking area and the necking area using the direct measurement method and the indirect measurement method, respectively. All thickness strain data have been filtered using the RANSAC algorithm before plotting the R-value curves. It can be seen from the figures that the two measurement curves of the non-necking area are very close. The R-values obtained by fitting the two curves with least squares are 0.4895 and 0.4925, respectively. However, the data taken in the necking area shows the curves of the two measurement methods are different in the necking stage before fracture. The thickness strain of the direct measurement method increases more rapidly than that resulting from the indirect measurement method; thus, the R-value becomes smaller. The R-value curve change of this aluminum alloy 6061 based on direct measurement is similar to that of DP980. The constant volume assumption is no longer followed in the necking stage. The possible reason for this is that the crystal structure of the metal material is dislocated during the necking process.

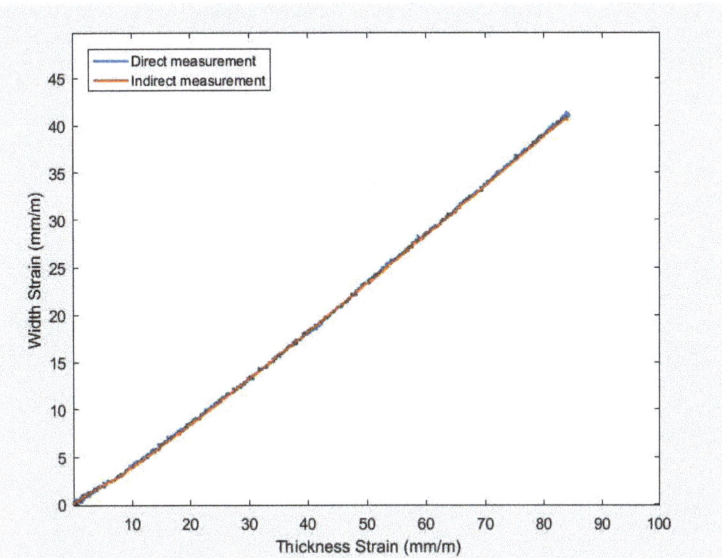

Figure 16. R-value curves of aluminum alloy 6061 in the non-necking area.

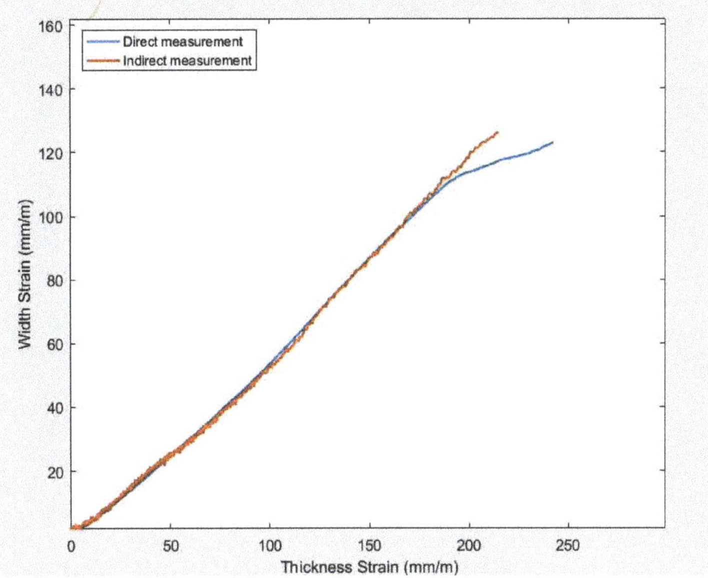

Figure 17. R-value curves of aluminum alloy 6061 in the necking area.

3.4.2. Polymer

The R-value of the polymer material is measured in this section. A 2.5 mm thick dog-bone specimen was used for uniaxial tensile experiments. Since the polymer material is a typical brittle material, the material does not neck during the tensile test. Any rectangular area of 8 × 8 pixels on the material surface can be selected for R-value calculation. Figure 18 shows the R-value curves of the two measurement methods for this material. It can be found that the measurement results of the two methods are linearly increasing, but the growth rate is quite different. The R-value of the direct measurement is 0.3101, while the indirect measurement result is 0.2499. This states that the polymer material does not entirely follow the constant volume assumption in the tensile test. One possible explanation is the constant volume assumption is based on the crystal structure of metal materials, while polymers are composed of polymer compounds with complex structures and no longer follow the deformation conditions of metal materials.

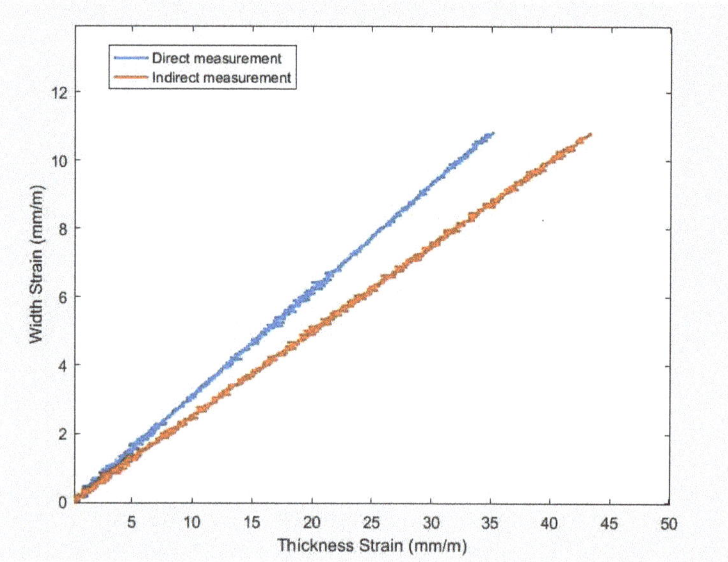

Figure 18. R-value curves of polymer.

4. Conclusions

A new R-value measurement method based on the use of two stereo-DIC systems has been presented in this paper. This method provides a new possibility for direct and simultaneous measurement of strains in the length, width, and thickness directions, thus allowing for direct measurement of the R-value. During the measurement process, it was found that DP980 and aluminum alloy 6061 do not follow the constant volume assumption in the necking stage. It was also found that a polymer material does not follow this assumption during the entire tensile test. The test results have certain significance for the research of different materials. Compared with the traditional R-value measurement method, this method has the advantages of direct measurement without assumption. The constant volume assumption has certain limitations, especially in the necking stage of material tensile deformation and polymer tensile deformation. Thus, the R-value obtained by the direct measurement method has more research value.

A main challenge to measuring strains in the third direction is that the strain data in this direction has more noise than other two strains due to the small value of the thickness. A standard DIC algorithm cannot be applied for such a measurement without using a special noise-removing algorithm, such as RANSAC presented in this paper. Some application limitations include the cost of four cameras and the large space requirement of the front and rear camera groups. Nevertheless, this paper provides an alternative method that directly measures R-value. The results for direct R-value measurement, as well as the comparison with the indirect R-value method, show that the method and the algorithm proposed is feasible and effective.

This paper has directly and indirectly measured the R-value of different materials, compared the results of the two measurement methods, and shown the accuracy of the direct measurement method. However, these measurement results are based on uniaxial tensile tests. Thus, it cannot completely reveal the reasons and limitations of the constant volume assumption. Future work may be necessary to study the microstructure of different materials before and after uniaxial tensile tests and determine the scope of application of the constant volume assumption through the analysis of the material microstructure.

Author Contributions: Conceptualization, S.F. and L.Y.; methodology, S.F., X.Z., B.Z.; validation, S.F., G.Z. and B.Z.; formal analysis, B.G.; investigation, G.Z.; resources, S.F., X.Z., B.Z.; data curation, S.F., G.Z.; writing—original draft preparation, S.F.; writing—review and editing, L.Y.; visualization, S.F., G.Z.; supervision, L.Y.; project administration, L.Y. All authors have read and agreed to the published version of the manuscript.

Funding: This research received no external funding.

Institutional Review Board Statement: Not applicable.

Informed Consent Statement: Not applicable.

Data Availability Statement: Data sharing not applicable.

Acknowledgments: The authors would like to express their sincere thanks to Bernard Sia, who made careful correction of the manuscript as well as the English.

Conflicts of Interest: The authors declare no conflict of interest.

References

1. ASTM. *E517–00. Standard Test Method for Plastic Strain Ratio R for Sheet Metal*; American Society for Testing and Materials: West Conshohocken, PA, USA, 2006.
2. Lankford, W.T.; Snyder, S.C.; Bausher, J.A. New criteria for predicting the press performance of deep drawing sheets. *Trans. Am. Soc. Met.* **1950**, *42*, 1197–1232.
3. Stickels, C.; Mould, P. The use of Young's modulus for predicting the plastic-strain ratio of low-carbon steel sheets. *Metall. Mater. Trans. B* **1970**, *1*, 1303–1312. [CrossRef]
4. Ghosh, M.; Miroux, A.; Kestens, L. Correlating r-value and through thickness texture in Al–Mg–Si alloy sheets. *J. Alloys Compd.* **2015**, *619*, 585–591. [CrossRef]
5. Moreau, A.; Lévesque, D.; Lord, M.; Dubois, M.; Monchalin, J.-P.; Padioleau, C.; Bussière, J. On-line measurement of texture, thickness and plastic strain ratio using laser-ultrasound resonance spectroscopy. *Ultrasonics* **2002**, *40*, 1047–1056. [CrossRef]
6. Huang, M.; Cheng, H. Determination of all elastic and plastic parameters for sheets of cubic metals only by uniaxial tension tests. *Eur. J. Mech. A/Solids* **2015**, *49*, 539–547. [CrossRef]
7. Bo, W.; Chen, X.-H.; Pan, F.-S.; Mao, J.-J.; Yong, F. Effects of cold rolling and heat treatment on microstructure and mechanical properties of AA 5052 aluminum alloy. *Trans. Nonferrous Met. Soc. China* **2015**, *25*, 2481–2489.
8. Seiner, H.; Bodnárová, L.; Sedlák, P.; Janeček, M.; Srba, O.; Král, R.; Landa, M. Application of ultrasonic methods to determine elastic anisotropy of polycrystalline copper processed by equal-channel angular pressing. *Acta Mater.* **2010**, *58*, 235–247. [CrossRef]
9. Mao, W.; Ma, Q.; Feng, H.; Chen, C.; Li, J. On-line determination technology on Lankford parameter of deep drawing aluminum sheets. *Chin. J. Nonferrous Met.* **2006**, *16*, 1149.
10. Huh, J.; Kim, Y.; Huh, H. Measurement of R-values at intermediate strain rates using a digital speckle extensometry. In *Dynamic Behavior of Materials*; Springer: Cham, Switzerland, 2011; Volume 1, pp. 317–322.
11. Esmaeilizadeh, R.; Khalili, K.; Mohammadsadeghi, B.; Arabi, H. Simulated and experimental investigation of stretch sheet forming of commercial AA1200 aluminum alloy. *Trans. Nonferrous Met. Soc. China* **2014**, *24*, 484–490. [CrossRef]
12. Xie, X.; Li, J.; Sia, B.; Bai, T.; Siebert, T.; Yang, L. An experimental validation of volume conservation assumption for aluminum alloy sheet metal using digital image correlation method. *J. Strain Anal. Eng. Des.* **2017**, *52*, 24–29. [CrossRef]
13. Chu, T.; Ranson, W.; Sutton, M.A. Applications of digital-image-correlation techniques to experimental mechanics. *Exp. Mech.* **1985**, *25*, 232–244. [CrossRef]
14. Chen, X.; Xu, N.; Yang, L.; Xiang, D. High temperature displacement and strain measurement using a monochromatic light illuminated stereo digital image correlation system. *Meas. Sci. Technol.* **2012**, *23*, 125603. [CrossRef]
15. Helm, J.D.; McNeill, S.R.; Sutton, M.A. Improved three-dimensional image correlation for surface displacement measurement. *Opt. Eng.* **1996**, *35*, 1911–1920. [CrossRef]
16. Chen, F.; Chen, X.; Xie, X.; Feng, X.; Yang, L. Full-field 3D measurement using multi-camera digital image correlation system. *Opt. Lasers Eng.* **2013**, *51*, 1044–1052. [CrossRef]
17. Li, J.; Dan, X.; Xu, W.; Wang, Y.; Yang, G.; Yang, L. 3D digital image correlation using single color camera pseudo-stereo system. *Opt. Laser Technol.* **2017**, *95*, 1–7. [CrossRef]

18. Xie, X.; Li, J.; Zhang, B.; Sia, B.; Yang, L. Direct measurement of R value for aluminum alloy sheet metal using digital image correlation. In *International Digital Imaging Correlation Society*; Springer: Cham, Switzerland, 2017; pp. 89–94.
19. Li, J.; Xie, X.; Yang, G.; Zhang, B.; Siebert, T.; Yang, L. Whole-field thickness strain measurement using multiple camera digital image correlation system. *Opt. Lasers Eng.* **2017**, *90*, 19–25. [CrossRef]
20. Vendroux, G.; Knauss, W. Submicron deformation field measurements: Part 2. Improved digital image correlation. *Exp. Mech.* **1998**, *38*, 86–92. [CrossRef]
21. Zhang, Z. A flexible new technique for camera calibration. *IEEE Trans. Pattern Anal. Mach. Intell.* **2000**, *22*, 1330–1334. [CrossRef]
22. Fischler, M.A.; Bolles, R.C. Random sample consensus: A paradigm for model fitting with applications to image analysis and automated cartography. *Commun. ACM* **1981**, *24*, 381–395. [CrossRef]
23. Song, J.; Yang, J.; Liu, F.; Lu, K. High temperature strain measurement method by combining digital image correlation of laser speckle and improved RANSAC smoothing algorithm. *Opt. Lasers Eng.* **2018**, *111*, 8–18. [CrossRef]
24. Li, Q.; Zhang, H.; Chen, F.; Xu, D.; Sui, D.; Cui, Z. Study on the plastic anisotropy of advanced high strength steel sheet: Experiments and microstructure-based crystal plasticity modeling. *Int. J. Mech. Sci.* **2020**, *176*, 105569. [CrossRef]

Article

A Novel Machine-Learning-Based Procedure to Determine the Surface Finish Quality of Titanium Alloy Parts Obtained by Heat Assisted Single Point Incremental Forming

Fernando Bautista-Monsalve [1,2], Francisco García-Sevilla [1,2], Valentín Miguel [1,2,*], Jesús Naranjo [1,2] and María Carmen Manjabacas [1,2]

[1] High Technical School of Industrial Engineers, University of Castile—La Mancha, Paseo de los Estudiantes s/n, 02008 Albacete, Spain; Fernando.Bautista1@alu.uclm.es (F.B.-M.); Francisco.Garcia@uclm.es (F.G.-S.); Jesus.Naranjo@uclm.es (J.N.); Mcarmen.Manjabacas@uclm.es (M.C.M.)

[2] Regional Development Institute, Material Science and Engineering, University of Castilla—La Mancha, 02006 Albacete, Spain

* Correspondence: valentin.miguel@uclm.es; Tel.: +34-96-7599200

Abstract: Single point incremental forming (SPIF) is a cheap and flexible sheet metal forming process for rapid manufacturing of complex geometries. Additionally, it is important for engineers to measure the surface finish of work pieces to assess their quality and performance. In this paper, a predictive model based on machine learning and computer vision was developed to estimate arithmetic mean surface roughness (Ra) and maximum peak to valley height (Rz) of Ti6Al4V parts obtained by SPIF. An image database was prepared to train different classification algorithms in accordance with a supervised learning approach. A speeded up robust feature (SURF) detector was used to obtain visual vocabulary so that the classifiers are able to group the photographs into classes. The experimental results indicated that the proposed predictive method shows great potential to determine the surface quality, as classifiers based on a support vector machine with a polynomial kernel are suitable for this purpose.

Keywords: HA-SPIF; surface finish; machine learning; Ti6Al4V

Citation: Bautista-Monsalve, F.; García-Sevilla, F.; Miguel, V.; Naranjo, J.; Manjabacas, M.C. A Novel Machine-Learning-Based Procedure to Determine the Surface Finish Quality of Titanium Alloy Parts Obtained by Heat Assisted Single Point Incremental Forming. *Metals* **2021**, *11*, 1287. https://doi.org/10.3390/met11081287

Academic Editor: Umberto Prisco

Received: 21 June 2021
Accepted: 13 August 2021
Published: 16 August 2021

Publisher's Note: MDPI stays neutral with regard to jurisdictional claims in published maps and institutional affiliations.

1. Introduction

Market requirements, along with the development of the industrial sector, have led to parts being demanded under restrictive specifications. As a result, surface finish has emerged as a key aspect of many products in which this feature is critical.

Single Point Incremental Forming (SPIF) is a rapid manufacturing technique based on the localized plastic deformation of sheet metal by a hemispherical tool [1]. For that, the sheet is fixed to a rigid frame and the deformation is produced, moving the tool according to an helicoidal path and establishing a pitch value to feed the operation. The method has generated considerable research interest because it overcomes some of the disadvantages of conventional forming. The SPIF process is characterized by its ability to economically develop complex geometries by means of reduced tooling costs if short series or prototypes are being produced, notably in biomedical and aeronautical applications [2–4]. The process has been applied to commercial copper, some aluminum alloys, and high formability mild steel. Some application with stainless steel AISI 304 has been also reported [5]. With regard to titanium alloys, research is mainly recent, and it supposes that their application has the most potential. However, titanium alloys, particularly Ti6Al4V, have a poor formability at room temperature which classifies them into the hard-to-form materials group. Thus, it is necessary to heat the sheet to carry out the SPIF process, also called heat assistant SPIF or HA-SPIF. Consequently, substantial efforts have been undertaken in recent years to develop heating methods. Götmann et al. [6] worked on the process of dynamic heating by laser, previously proposed by Duflou et al. [7]. Fan et al. [8] developed a heating procedure based on Joule heating by electric power from the tool to the sheet, which has also been adopted by other authors [9]. Ambrogio et al. [10] designed an external heating system using electric bands. Similar systems have been applied by Naranjo et al. [11] and Ortiz et al. [12]. There have been attempts to improve formability using the local heating generated between the tool friction on the sheet metal when it rotates at high speed [13]. This procedure is known as Stir-SPIF. Finally, some methods mixing friction and external sources of heat can be found in literature [14].

Despite the improvement reached by HA-SPIF methods, the obtained surface finishes are low quality. There have been some efforts to improve the surface finish after the SPIF process trying to reduce the friction tool sheet without lubricant [15] or considering the effect of lubricant on the process [16], but with a few exceptions, the research has been focused on only room temperature and not on titanium alloys.

Additionally, the influence of surface quality on the operability of the parts promotes the constant development of measurement methods. In recent decades, this sector has undergone successive updates in order to obtain a reliable, precise, and flexible technique to measure surface finish.

Contact profilometry is one of the most common methods in surface finish measuring despite having several drawbacks. Its use is often restricted to controlled environments because it is highly sensitive to vibrations and the stylus can cause flaws on the surface of the part. These limitations, together with its slowness, render it an unsuitable method for measuring online surface roughness.

Optical non-contact methods arose as a necessary alternative to overcome these drawbacks. Of these techniques, the procedure developed by Moreas et al. [17] should be mentioned. In this work, an optical microscope was built to capture images and topography was studied using the principle of triangulation. From another perspective, in the research conducted by Lehmann [18], interferometry techniques were employed in order to analyze the statistical properties of speckle patterns to obtain significant surface roughness parameters. However, non-destructive optical methods do not always achieve the level of precision provided by contact profilometry and samples must be adapted for each of them. Particularly, optical microscopes equipped with profilometer function could be appropriate with higher accuracy in the measurement of surface roughness, but the measuring field is too small and the sample must be split from the part. The dominant key issues for future research include systematic process planning, tool-path optimization, compensating unwanted deformations, advanced feature recognition techniques, forming reliable and robust communication interfaces, flexible hardware, adaptive control methods, the ability to form newer materials and a wider range of materials, complex shapes with ease and accuracy, and removing barriers on sizes of components.

Artificial intelligence has become very popular in this field because machine learning techniques are capable of developing predictive models that provide better results than analytical ones. Machine learning algorithms have been shown to help reduce the number of experiments required to establish a correlation between surface roughness and manufacturing parameters with great effectiveness.

Thus, machine learning has been used to analyze the influence on surface finish of the parameters involved in the manufacturing process. The aim is to optimize the process by maximizing the number of suitable parts. In this regard, Abd et al. [19] developed a gradient boosting regression tree (GBRT) algorithm to establish a correlation between the operating parameters and surface quality of aluminum/stainless steel (Al/SUS) bimetal sheets obtained by SPIF. This research used artificial intelligence to expand on bimetal sheets the results obtained by Echrif and Hrairi [20] in single-layer sheets of aluminum alloy AA105-0.

The use of machine learning algorithms for measuring surface quality allows the limitations of traditional methods to be overcome. Models built with this methodology have proven to be accurate and able to adapt production speeds. In this context, the work by Mulay et al. [21] focuses on a feedforward backpropagation network that was implemented to predict the average surface roughness and maximum forming wall-angle of parts of aluminum alloy AA5052-H32 obtained by SPIF. This experiment showed that artificial neural networks are a promising tool with great economic benefits, greater predictability, and low simulation time.

The work developed by Abu-Mahfouz et al. [22] considered a classifier based on a support vector machine to obtain surface roughness from vibration signals collected during the milling of aluminum parts. In addition, this work analyzed the effect of different types of kernels and sets of features on the performance of the model. Their proposed classification model was trained with 32 different combinations of the milling process conditions and managed to predict the surface roughness with an accuracy of 81.25%.

Furthermore, the research by Koblar et al. [23] developed a computer vision method based on machine learning for surface roughness measurement. Decision trees were used to solve classification and regression problems on a database formed by 300 commutator images taken in 8-bit grayscale. This experiment offers new possibilities for evaluating surface quality and develops a suitable method for applying online roughness measurement without contact. However, the nature of the work calls for the use of a broad set of samples with respect to inputs based on manufacturing parameters.

The research conducted by Kurra et al. [24] developed predictive models for the average and the maximum surface roughness measurement (Ra and Rz) after the SPIF process based on different techniques: Artificial Neural Networks, Support Vector Regression, and Genetic Programming. In the training of these models, the effect of tool diameter, step depth, wall angle, feed rate, and lubricant type were taken into consideration as process variables. Each parameter was varied over three levels to study different combinations of manufacturing parameters and achieve a uniform distribution of roughness values. Roughness was thus analyzed from a technical viewpoint, so this method requires knowing in advance the conditions in which the parts were manufactured.

Other works, such as that by Lin et al. [25], have used deep neural networks, long short-term memory networks, and one-dimensional convolutional neural networks for the prediction of arithmetic mean roughness from vibration signals obtained in the milling process of S45C steel parts. Li et al. [26] compared the results from different classifiers to predict the surface roughness of polylactic acid intake flanges obtained by 3D printing. In both studies, Fast Fourier Transform (FFT) was used to extract the features' raw data and different techniques were evaluated to build an accurate predictive model.

The literature review showed that several studies have developed predictive models for surface quality measuring of work pieces produced by different processes. However, no research has focused on the application of machine learning techniques to predict the surface quality from images of pieces created by incremental forming, among other reasons, because of the large number of samples required for the model training and the amount of time needed to acquire the features. To fill this gap, the aim of this work was to develop a new methodology for the evaluation of surface quality using photographs of twenty-one parts formed by SPIF as inputs in machine learning classifiers. This procedure would allow the surface finish to be determined without the need to previously know the manufacturing conditions. In this way, surface quality could be studied in real time during the production process.

The remainder of the paper is organized as follows. Section 2 contains an introduction on machine learning and explains the experimental setup in detail. Section 3 describes the data preparation and the training of predictive models. Section 4 presents an evaluation of the predictive model performance and discusses the experimental results. Finally, Section 5 provides findings and future directions.

2. Materials and Methods

Machine learning is a subset of artificial intelligence, the purpose of which is to train computers so they are able to learn automatically through experience. Algorithms are used to build a predictive mathematical model capable of generalizing a behavior without being explicitly programmed [27,28]. In this way, the computer acquires the ability to recognize patterns in sample data to predict future unseen data. Machine learning methods are usually categorized in accordance with their strategy, data processing, and the objective of the application.

Firstly, in the supervised learning approach training, data are composed of a set of previous analyzed examples whose outputs are known. Using these observations as inputs, the computer can learn general rules and establish a correlation to obtain an inferred function, being able to determine the output for new data. On the other hand, human intervention is not required in unsupervised learning algorithms. In this case, a predictive model is trained with unseen data that have not been studied. As a result, it must identify a pattern without previous knowledge. Thirdly, reinforcement learning is a method built on the interaction with its environment. Feedback is used to develop the model through rewards and punishments. Systems learn by analyzing the behavior of their surroundings in response to their actions to maximize positive signals.

This paper focuses on supervised learning to build a predictive model to obtain arithmetic mean roughness from the images of a piece manufactured by incremental forming; as a result, labelled and categorized photographs are used as training data. To this end, the developed model must be capable of extracting characteristics from examples and achieving a relationship between image features and outputs in order to apply this knowledge on new unobserved data.

Having set the objective, the working methodology can be divided into two separate stages: the first phase, aimed at measuring the parts and creating a database that would be used to train machine learning algorithms; the second stage concerning the development and the validation of the classification model.

Experimental Details of Measuring Process of Workpieces

We measured 0.8 mm thick Ti6Al4V alloy parts obtained by a heat-assisted SPIF process to determine the 3D average and maximum roughness parameters (Sa and Sz). The corresponding waviness parameters (Wa and Wz) were also considered. The pieces consisted of cone trunks with a variable wall angle; the angle of the cones was continuously increasing with the forming depth in order to determine the SPIF formability of the alloy. WC and steel tools with a spherical tip of 10 and 12 mm were used and the processing temperature was within the range of 25–400 °C [11]. Parts were created in a Deckel Maho machining center, model DMC 835V (DMG MORI, CO., LTD.,Bielefeld, Germany). A hole frame was built to fasten the sheet. An electric suitably isolated mini-oven was specifically designed and fitted inside the support. The forming strategy adopted was a helical path with a pitch of 0.5 mm per revolution, a feed rate of 600 mm per min, and without any lubrication. To minimize any friction between tool and sheet, a counter-rotation speed of the tool was set at 30 rpm.

The sheet in virgin state always is covered by a TiO_2 layer that was not removed by a previous surface treatment, as it is usual in the SPIF for guaranteeing a same initial surface state in all tests.

The parts were specially selected to have a heterogeneous collection of surface finish values (Table 1). The roughness and waviness values reported in Table 1 correspond to measurements taken in intermediate positions of the rolling direction.

A profilometer Form Talysurf 50 and the Talymap software (6.0, Taylor Hobson Ltd., LE4 9JQ, Leicester, UK) were used to measure the roughness and to process the measuring data. An inductive gauge with a 2 μm radius was employed. A block diagram of the measurement procedure is shown in Figure 1. Measurements were carried out on surfaces of 5 mm × 2 mm along three different directions with respect to the rolling direction of the sheet, that is, 0°, 45°, and 90°. The measuring instrument provided an initial profile with some differences to the real surface because of the limitations of the contact profilometry, that is, the gauge radius. From the measured surface, the primary surface is obtained once its shape (the curvature in this case) is removed. Finally, by using a gaussian filter, the irregularities of the primary profile can be split up in the roughness and waviness profiles.

Table 1. Single point incremental forming parameters and surface finish values. The value of wall angle in brackets corresponds to the maximum angle formed. (Initial sheet thickness = 0.8 mm, pitch = 0.5 mm/rev, feed rate = 600 mm/min).

Test	Temp. (°C)	Angle (°)	Max. Depth (mm)	Tool (Diameter (mm)/ Material)	Ra (µm)	Rz (µm)	Wa (µm)	Wz (µm)
1	Room	Var (30)	11.4	12/WC	0.976	18.2	0.539	5.73
2	200	Var (40)	17.1	12/WC	0.561	9.87	0.356	2.48
3	300	Var (40)	18.1	12/WC	0.534	15.5	0.344	2.82
4	400	Var (50)	21.1	12/WC	0.521	9.95	0.544	3.43
5	400	40	18	12/WC	1.1	34.8	1.39	17.9
6	400	45	18	12/WC	1.37	34.5	1.65	22.6
7	Room	Var (30)	11.3	12/STEEL	0.704	13.7	0.716	6.22
8	200	Var (40)	15.6	12/STEEL	0.828	14.4	0.801	6.93
9	300	Var (40)	17.5	12/STEEL	1.11	26.9	1.39	13.1
10	400	Var (50)	20.2	12/STEEL	1.05	25.3	0.93	8.03
11	Room	Var (30)	12.5	10/WC	0.862	18.4	0.541	5
12	Room	Var (30)	12.5	10/WC	1.18	20.9	0.841	5.08
13	200	Var (40)	16.3	10/WC	0.913	20.1	0.583	5.53
14	200	Var (40)	16.3	10/WC	0.808	20.3	0.68	5.8
15	300	Var (50)	21.5	10/WC	0.775	37	0.569	5.31
16	400	Var (50)	22.1	10/WC	1.61	26.3	2.09	10.8
17	200	Var (40)	16.4	12/STEEL	1.35	31.9	0.946	7.08
18	300	Var (40)	19.1	12/STEEL	0.816	11.6	0.708	6.05
19	300	Var (40)	19.8	12/STEEL	1.45	30.7	0.699	8.66
20	Room	Var (30)	11.8	12/STEEL	1.48	23.4	0.668	5.48

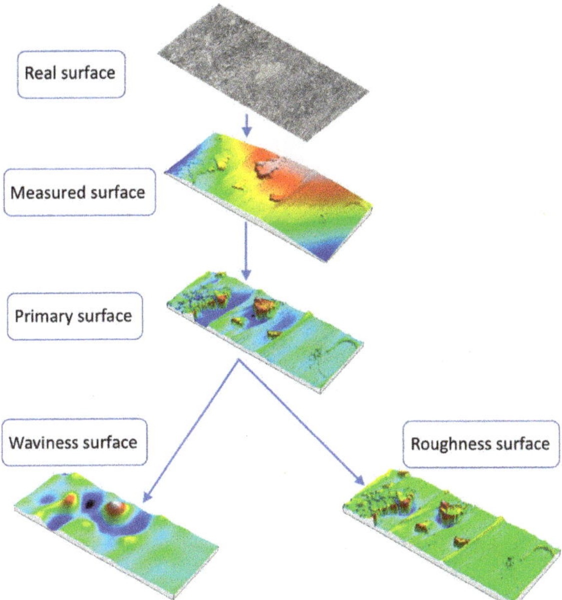

Figure 1. Measurement procedure.

According to the shape of the parts obtained by the SPIF process, it was considered appropriate to apply a polynomial of degree 7 to remove that shape and obtain the primary surface. Some filters were then studied in order to separate the roughness and waviness surfaces that comprise the primary surface. The Gaussian filter is currently the most widely used filtering tool in three-dimensional models because of its isotropic transmission and lack of phase shift according to the standards ISO 11562 and ASME B46. Nevertheless, the symmetry of the filter may cause problems in the edge treatment, which must be removed. Alternately, the spline filter is based on a matrix algorithm which prevents edge defects and is characterized by fast calculations and complex geometry filtering. Nevertheless, this filter has not been widely used in the measurement of 3D surfaces because of its severe anisotropic characteristics. Finally, the robust Gaussian filter used an iterative statistical method in order to improve the Gaussian filter response. This technique allows for a topographic reconstruction without being affected by accidental surface defects. However, it demands large amounts of computation time. Finally, a Gaussian filter, including an edge treatment, was applied by using the Talymap Gold software from Taylor Hobson (6.0, Taylor Hobson Ltd., Leicester, UK).

As a result of the application of the Gaussian filter, roughness and waviness three-dimensional profiles were obtained. These were then used to gather surface finish parameters to build a database.

Concurrently, an Olympus SZX7 (Olympus Corp., Tokyo, Japan) stereoscopic microscope was used to take photographs of the measured surfaces. The images, with dimensions of approximately 6.2 mm × 6 mm, were labelled in accordance with the direction and the position in which they were photographed. Additionally, every image was distributed in files associated with the part to which it belonged. As an example, Figure 2 shows the images taken from Part 5 in the rolling direction (0°). Although the references of the different areas are represented for a frustum with a variable wall angle, the part depicted in Figure 2 corresponds to a constant wall angle cone.

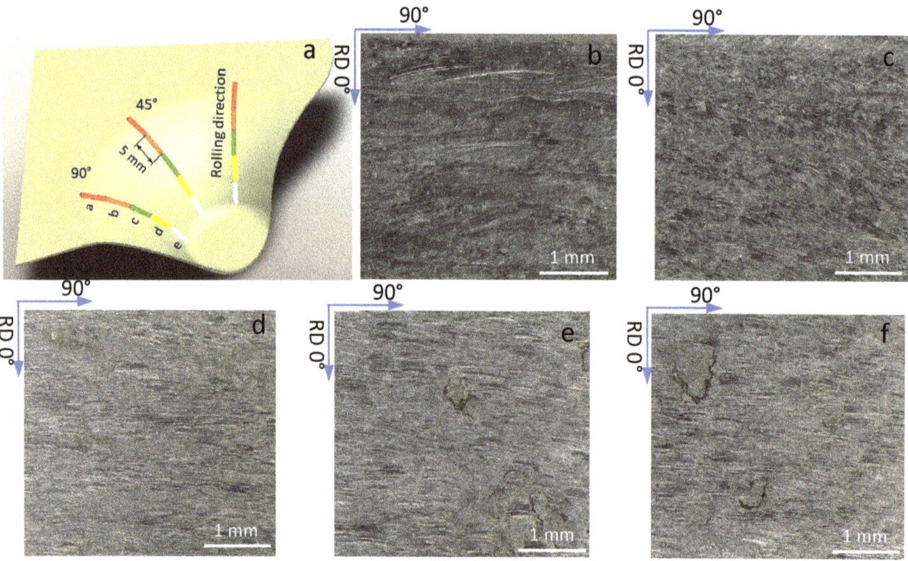

Figure 2. (**a**) Images corresponding to Part 5 (experiment 5) in the rolling direction at different depths; (**b–f**) correspond to Zones 1 to 5, respectively. The rolling direction, RD, corresponds to the vertical axis of the photographs.

3. Data Preparation and Model Training

Building a database is highly time consuming, not only to process the collected images, but also to train predictive models with them. For this purpose, it is necessary to previously know the functionality of the model.

Two types of problems can generally be distinguished when a supervised learning approach is used. First, classification questions are intended to categorize unknown data into known discrete ranges named classes, while the aim of regression algorithms is to predict a continuous value.

Initially, this work seeks to categorize images into several classes according to their surface roughness. Figure 3 depicts a block diagram where data processing and the training procedure are shown in detail. At this second stage, the classification learner from MATLAB software (2019b, MathWorks, Natick, MA, USA) was used.

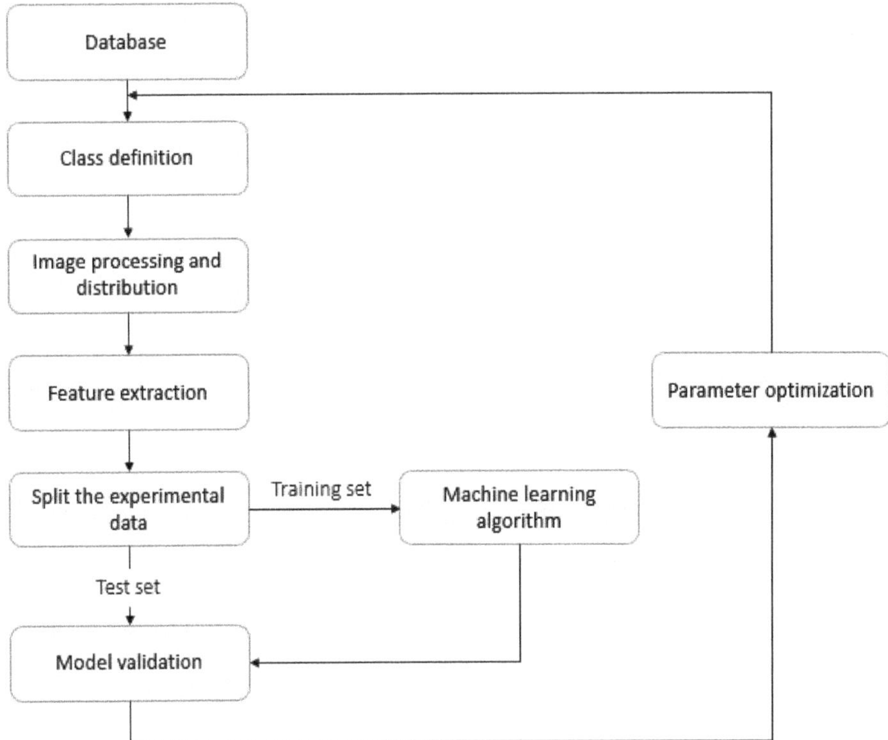

Figure 3. Data processing and model training.

Firstly, classes must be defined. Thus, the intervals for the arithmetic roughness, Ra, were class A [0.495, 0.799] μm, class B (0.799, 1.11] μm, and class C (1.11, 2.81] μm for three different class definitions. In this endeavor, it is necessary to collect the surface quality values from the name with which each image has been identified. In this way, it will be possible to set optimum roughness ranges to avoid imbalance in classes. As the number of groups increases, the accuracy required will be greater. Therefore, work begins on three classes. In accordance with the database used, trained models will only be able to solve classification tasks, whereas if we wanted to address regression tasks, a much larger number of images would be required.

Secondly, the photographs were processed and distributed in the specified classes. The aim of image processing is to maximize the picture features to facilitate their classification.

Coloring or dimensions are some of the characteristics that were analyzed. According to the literature, machine learning algorithms based on computer vision demand a large number of samples. For this reason, the possibility of splitting images into the prediction of average values was studied in order to increase the database.

The image characteristics were then extracted by using a speeded up robust feature (SURF) detector (2019b, MathWorks, MA, USA) in order to develop a visual vocabulary. The number of attributes was modified to optimize the accuracy of the predictive model. In addition, following the supervised learning method, the name of the classes was considered as another feature.

Next, the classification learner, included in Matlab tool boxes (2019b, MathWorks, MA, USA) was invoked, a software tool capable of training supervised learning algorithms from the table of characteristics and assessing the results obtained. A portion of the feature table created for training is shown in Table 2, where the class column can be seen.

Table 2. Part of the feature table used by the classification learner.

Image Number	Image Features 1	Image Features 2	Image Features 3	Image Features 4	Class
1	0.0232	0.0266	0.0598	0.0730	A
2	0.0300	0.0233	0.0566	0.0933	A
3	0.0149	0.0298	0.0410	0.0298	A
4	0.0517	0.0574	0.0345	0.0345	A
5	0.0101	0.0236	0.0608	0.0642	A
6	0.0667	0.0500	0.0083	0.0917	A
7	0.0416	0.0740	0.0416	0.0694	A
8	0.0091	0.0000	0.0091	0.0000	A
9	0.0270	0.0420	0.0480	0.0960	A
10	0.0509	0.0477	0.0477	0.0668	A

The support vector machine (SVM) is a supervised learning algorithm applied in countless fields to solve classification and regression problems [29]. Furthermore, following the literature [22,24,26], it has proven to be an effective technique to classify the surface roughness using higher dimensional data.

The operating principle of this classifier is based on finding an optimal hyperplane ensuring the best separation between classes, that is to say, to maximize the margin between the boundary function and the closest samples. The separation hyperplane can be defined by Equation (1), depending on its orthogonal vector w and intersection coefficient b and where x refers to the vector coordinates. Symbol T represents the transpose matrix-vector.

$$w^T \cdot x_i + b = 0 \tag{1}$$

In mathematical terms, this question becomes a quadratic programming optimization with the objective of minimizing the named margin inverse function (Equation (2)). The second term of Equation (2) represents the action of adjustment variables, ξ_i, where the C is a balancing factor.

$$\Phi(x) = \frac{1}{2}||w||^2 + C \cdot \sum_{i=0}^{n} \xi_i \tag{2}$$

In our view, the distribution of the samples as well as the number of classes demands this issue be addressed as a higher dimensional classification with non-linear separable data. To tackle this challenge, the original area must be mapped into a higher dimensional space in order to enable developing an appropriate separation function in the new conditions. This transformation is achieved with the help of kernel functions and its effect on the problem is reflected in the SVM dual formulation (Equation (3)), where the Lagrangian is utilized to solve the quadratic programming problem and the coefficients α reference the Lagrange multipliers.

$$\Theta(\alpha) = -\frac{1}{2} \cdot \sum_i \sum_j \alpha_i \alpha_j y_i y_j \cdot K(x_i x_j) - \sum_i \alpha_i \tag{3}$$

Kernel functions are chosen according to the nature of the problem. The suitability of different types in relation to the surface roughness prediction was discussed in the work by Abu-Mahfouz et al. [22], where a linear kernel was shown to be the most promising function. Nevertheless, all Kernel functions implemented in MATLAB® [30] were tested in order to find that one leading the best results, i.e., the highest precision.

Finally, the validation procedure depends on the database size. In this work, two techniques were used to avoid overfitting: holdout validation and cross validation. In holdout validation, a portion of data is selected to train the predictive model, whereas the remaining samples are used as a test set. For this reason, holdout validation is suitable for large data sets. On the other hand, cross validation is characterized by the splitting of the initial samples into folds. Each block is trained with the observations not belonging to the fold and is validated with the remaining images. Test error is calculated as an average of the mistakes in each fold. Therefore, according to the number of classes and images, different validation methods were used.

4. Results and Discussion

Considering only the right answers in measuring the effectiveness of the model may lead to an erroneous perception of the machine behavior. For this reason, a confusion matrix was used to assess the performance of the supervised learning algorithms. In contrast to other metrics, a confusion matrix is capable of distinguishing between different types of errors. It is a square matrix where each row represents the actual classes, while each column refers to predicted classes. As a result, its main diagonal reflects the correct predictions, and the rest of its cells represent misclassifications.

In addition to the accuracy of the classifiers, two other measures are provided by a matrix confusion, namely precision and recall. Precision is used to obtain the percentage of correct predictions in every class, meaning the degree of reliability, while recall is used to represent the fraction of samples which were correctly recognized, that is, the model's detection capability. Both measures were calculated using Equations (4) and (5). In these equations, TP (True Positive) refers the number of predictions where the classifier correctly predicts the positive class as positive, TN (True negative) indicates the number of predictions where the classifier correctly predicts the negative class as negative, FP (False Positive) depicts the samples incorrectly associated with a class, and FN (False Negative) represents the experiments belonging to a specific class that were wrongly labelled.

$$\text{Precision} = \frac{\text{TP}}{\text{TP} + \text{FP}} \tag{4}$$

$$\text{Recall} = \frac{\text{TP}}{\text{TP} + \text{FN}} \tag{5}$$

Combining both previous indicators, the F-score value can be obtained according to the Equation (6), which evaluates the harmonic mean of true positives and true negative cases.

$$\text{F} = \frac{2 \cdot \text{Precision} \cdot \text{Recall}}{\text{Precision} + \text{Recall}} \tag{6}$$

As discussed, the approach consisted of developing an effective model to predict the average surface roughness and to extrapolate the results to complementary variables. To this end, 153 simulations were carried out based on the procedure indicated in Figure 3. This led us to optimize the process parameters and maximize the accuracy of the predictive model. All the results are set out below, in a detailed description of the optimized model.

The model was trained with images obtained in JPG format with dimensions of 1148 × 1076 pixels, which were converted into RGB. These pictures were not split, since it has been proven that increasing the number of training images does not justify the deviation between the measured roughness and the value of the fragmented photograph. Additionally, features were extracted in each image through a SURF object recognition. At this point, the appropriate number of features was analyzed to optimize the accuracy of the model. Hence, models were trained using 2000, 1500, 1000, and 500 attributes, among other values.

In Table 3, some training tests are selected from the whole probes carried out. Particularly, those classified using Medium Gaussian SVM algorithm with three classes and using a photograph division from 1 into 4 (2 × 2) are shown. The number of features is correlated with the difficulty of classifying the photograph correctly, that is, the higher number of features, the more demanding requirements must be fulfilled. Thus, as it can be appreciated, the accuracy obtained is lower as the number of features is higher.

Table 3. Comparison between the number of features extracted and the model accuracy with a medium Gaussian SVM classifier.

Model	Number of Features	Classifier	Number of Classes	Photograph Fragmentation	F-Score
91	2000	Medium Gaussian SVM	3	2 × 2	0.54
55	1500	Medium Gaussian SVM	3	2 × 2	0.57
57	1000	Medium Gaussian SVM	3	2 × 2	0.57
59	750	Medium Gaussian SVM	3	2 × 2	0.58
-	<500	Medium Gaussian SVM	3	2 × 2	>0.60

Finally, an initial value of 500 features was selected. In such a way, a thorough analysis could be performed to assess the impact of each characteristic on the prediction capacity, and thus counterproductive features could be removed.

Under those conditions, all classifiers were trained. The supervised learning algorithms used to train the different models are shown in Figure 4.

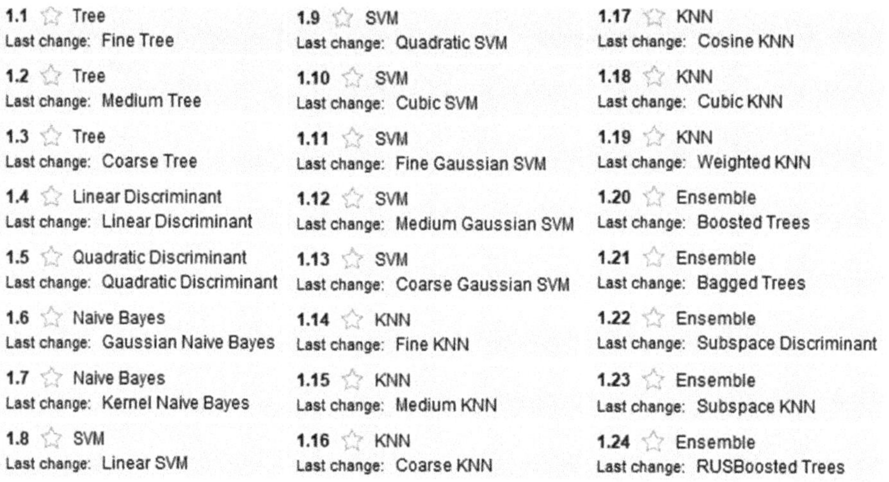

1.1 ☆ Tree
Last change: Fine Tree

1.2 ☆ Tree
Last change: Medium Tree

1.3 ☆ Tree
Last change: Coarse Tree

1.4 ☆ Linear Discriminant
Last change: Linear Discriminant

1.5 ☆ Quadratic Discriminant
Last change: Quadratic Discriminant

1.6 ☆ Naive Bayes
Last change: Gaussian Naive Bayes

1.7 ☆ Naive Bayes
Last change: Kernel Naive Bayes

1.8 ☆ SVM
Last change: Linear SVM

1.9 ☆ SVM
Last change: Quadratic SVM

1.10 ☆ SVM
Last change: Cubic SVM

1.11 ☆ SVM
Last change: Fine Gaussian SVM

1.12 ☆ SVM
Last change: Medium Gaussian SVM

1.13 ☆ SVM
Last change: Coarse Gaussian SVM

1.14 ☆ KNN
Last change: Fine KNN

1.15 ☆ KNN
Last change: Medium KNN

1.16 ☆ KNN
Last change: Coarse KNN

1.17 ☆ KNN
Last change: Cosine KNN

1.18 ☆ KNN
Last change: Cubic KNN

1.19 ☆ KNN
Last change: Weighted KNN

1.20 ☆ Ensemble
Last change: Boosted Trees

1.21 ☆ Ensemble
Last change: Bagged Trees

1.22 ☆ Ensemble
Last change: Subspace Discriminant

1.23 ☆ Ensemble
Last change: Subspace KNN

1.24 ☆ Ensemble
Last change: RUSBoosted Trees

Figure 4. Compilation of classifiers used.

Regarding the outcome evaluation, given the limited number of training data available, a fivefold cross validation was used. Consequently, the best performance was obtained in

Model 141 with a support vector machine with a quadratic kernel, the confusion matrix of which is shown in Figure 5. This model achieved an accuracy of 69.4%.

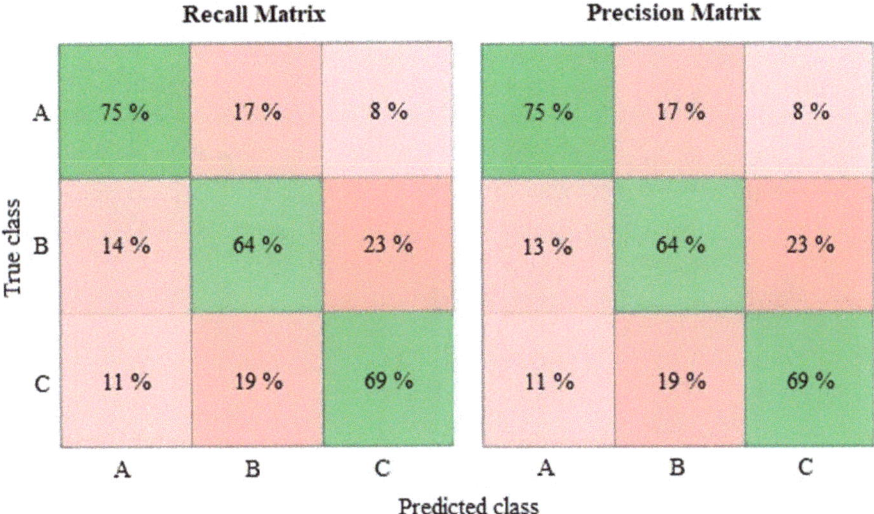

Figure 5. Confusion matrices of Predictive Model 141 for average surface roughness, Ra.

According to the recall matrix, the confusion is greater when distinguishing images belong to Classes B and C. Based on percentages, 23% of Class B images were incorrectly associated with Class C, whereas 19% of Class C samples were wrongly linked to Class B. Therefore, the predictive model displayed a better detection capability for roughness values in the first interval (Class A). Additionally, the precision matrix reflects a similar performance to that of the recall matrix, and so Class A presents the greatest reliability.

As can be seen from Figure 5, the intermediate class presented less reliability and recall with regard to the other ones, given it was more difficult to extract specific features of this class. This was a recurring issue that, among other problems, may be associated with a narrow roughness range and an insufficient database.

On the one hand, non-uniform distribution of roughness values resulted in a reduction in inequalities between categories. As a result, there are fewer specific characteristics that lead to further confusion in the classification. On the other hand, the inaccuracy of measuring instruments gains importance, since a small error, around 10 μm, might derive in an incorrect classification.

Once the average surface roughness had been studied, new predictive models were built to assess the methodology on complementary parameters, such as the maximum peak to valley height and the arithmetic mean waviness. Of these, we should highlight the following model trained to predict the average primary surface.

Model 169 was trained and validated using the same settings as the previous one. In this case, an accuracy of 80% was reached with a support vector machine with a cubic kernel. Figure 6 depicts the model confusion matrix.

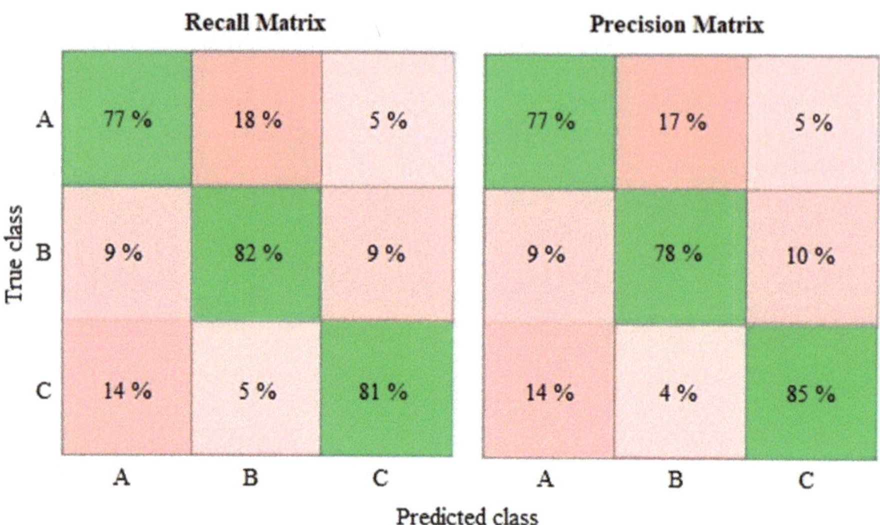

Figure 6. Confusion matrices of Predictive Model 169 for the average primary surface measurement.

As can be seen, recall and precision are approximately 80% in all classes. This improvement can be explained by the greater coherence between photographs and parameters, given that the images used represent an overlapping between roughness and waviness. In accordance with the better model performance, not only could the effectiveness of machine learning algorithms in this field be proved, but also, the application of filters on samples is considered for further work.

After running 169 simulations, changing processing parameters to build predictive models for surface quality measuring, the best results were those reported in Table 4.

Table 4. Results of the models developed. Intervals for the arithmetic roughness, Ra, are: 3 classes—class A [0.495, 0.799] μm; class B (0.799, 1.11] μm; class C (1.11, 2.81] μm; 4 classes—class A [0.407, 0.7] μm; class B (0.7, 0.95] μm; class C (0.95, 1.15] μm; class D: (1.15, 2.81] μm.

Model	Variable	Classes	Classifier	Kernel	F-Score	Validation	Training Time (s)
141	Ra	3	SVM	Quadratic	0.69	Cross	1.7324
166	Ra	4	KNN	Fine	0.57	Cross	0.9927
163	Wa	3	SVM	Quadratic	0.72	Holdout	0.3857
164	Wa	4	SVM	Cubic	0.61	Holdout	1.0731
167	Rz	3	SVM	Cubic	0.56	Cross	1.6442
168	Wz	3	KNN	Cosine	0.63	Cross	0.9684
169	Pa	3	SVM	Cubic	0.80	Holdout	0.3651

The active participation of the support vector machine as main classifier supports the findings obtained in the literature [22,24,26]. Additionally, it is clear that this methodology is more appropriate for average parameter measuring.

The development of this work was subject to limitations that explain the results obtained, such as the small number of training images, considering the difficulty of the feature extraction in these types of images, the narrowness of the working range and the non-uniform distribution of the roughness values. Some of these are now discussed with a view to achieving better predictive models in future works. Firstly, the working range must be taken into account. Because of the narrowness of the average surface roughness interval, specific feature extraction for each class becomes a more complex task. Moreover, a strong adhesive mechanism existed in the tribology system modelled. This involved a non-uniform surface in all the cases considered herein. Thus, the surfaces presented scattered areas with removal of material and ploughing with material accumulations, besides the typical erosive abrasive with no clear tendency across the different SPIF process conditions (Figure 7). Figure 7 depicts the surface of a part measured in a 45° direction with respect to the rolling one, in different areas. In Figure 7a, the plastic deformation caused by the overlap between passes of the tool is marked as a regular macro-ploughing effect, indicated by arrows in the image. Figure 7b shows some hints of peeling pits (some of which are marked inside circles), while in Figure 7d, some re-adhesion of previous removed material can be observed. All these phenomena contrast with an almost uniform worn surface (Figure 7c) with only a micro-ploughing phenomenon with some isolated areas of peeling pits.

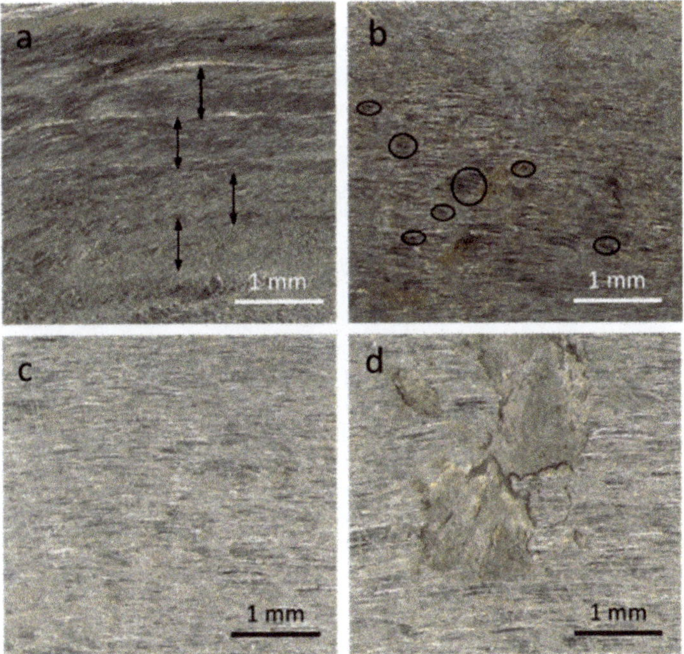

Figure 7. Surfaces of a part obtained by SPIF process at 400 °C with a WC tool of 12 mm diameter; surfaces correspond to 45° with respect to the rolling direction and from different zones of the part; (**a–d**) photographs correspond to zones 1 to 4 according to the reference of Figure 2. Zone 4 is the deepest.

Only one major adhesive mechanism was confirmed with the temperature, but its influence on the part does not always occur in the same area (Figure 7a–d). Some evidence of this can be observed in Figure 8, where the re-adhesion process is clearly noticeable in different areas of the images (marked areas). Moreover, the peeling–adhesion phenomena can dominate the topography of a study area, as can be seen in Figure 8a.

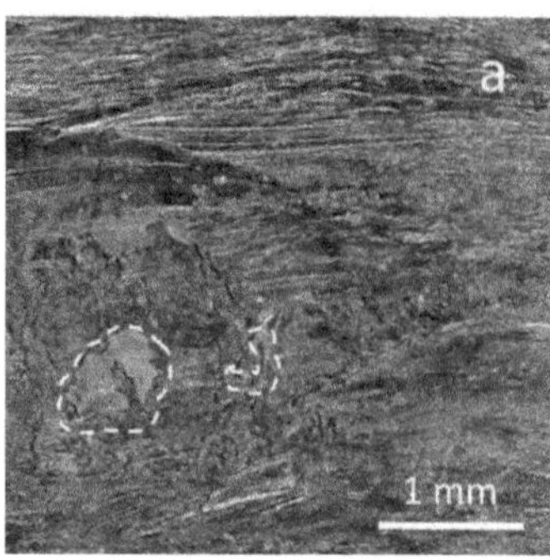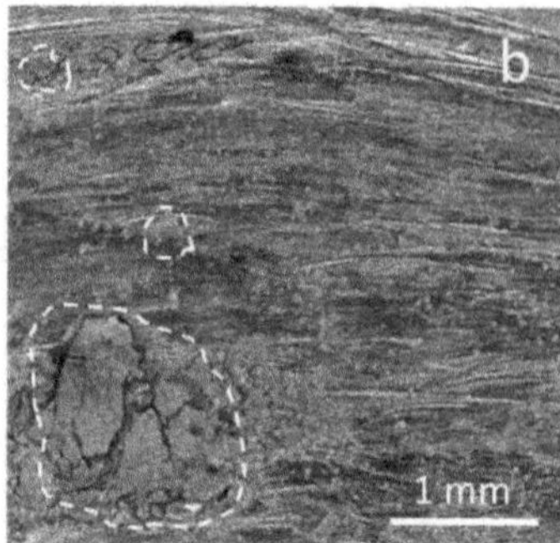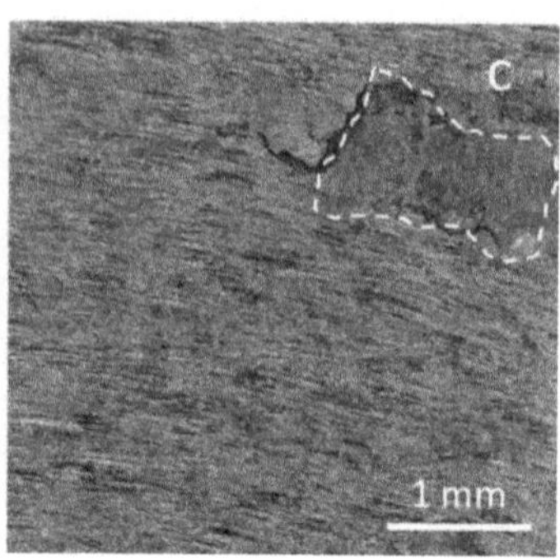

Figure 8. (**a**) Surface of a part obtained by SPIF process at 300 °C with WC tool of a diameter of 12 mm; inner area 2 in the rolling direction. (**b**) Surface of a part obtained at 300 °C with steel of 12 mm diameter; inner area 2 in the rolling direction. (**c**) Surface of a part obtained at 400 °C with WC tool of 12 mm diameter; inner area 5 in a 45° rolling direction. Inner areas according to the reference of Figure 2.

Increasing the number of observations, and thus the database size, would allow us not only to develop a margin of separation between classes, but also to define each category more accurately. While our work sets no margin, studies such as that conducted by Abu-Mahfouz et al. [22] established a distance of up to 0.18 μm between adjacent classes. As a result, classifiers based on machine vector supports using polynomial kernels achieved an 81.25% success rate.

In short, non-uniform distribution of training data demands a greater level of accuracy in both the measurement and the photography. The rigor required could be reached through the following steps:

- Noise reduction: numerous measurements of the same area could allow more stable values to be obtained.
- Modification of photography technique: the strategy proposed by Moreas and Bilstein [31] emphasized the relationship between the surface roughness and the peaks and valleys area. In this sense, an improvement to take into account in future work would be to use oblique illumination in an optical microscope.

Secondly, another limitation is the dataset size. The small amount of training data is a common problem in studies using machine learning algorithms. Following the literature, the number of samples required varies according to the purpose of the application. Whereas experiments using processing parameters utilize considerably fewer than a hundred observations, those using images to predict variables demand hundreds of photographs. This is evidenced by the work by Koblar et al. [23], in which 300 pictures were used to determine whether parts were suitable for commercialization in accordance with the difference between the highest peak and the lowest valley.

It is also necessary to emphasize the importance of a uniform distribution of images that allows us to use a suitable classification strategy without class imbalance. It only and exclusively depends on the manufacturing system as, in many cases, these conditions affect the surface finish of the parts in range and typology.

Finally, together with these limitations, the nature of the images must be considered. Roughness and waviness are overlapped in each photograph, so it is important to develop a filtering methodology in order to ensure that every picture faithfully represents the measured value.

5. Conclusions

In this paper, models based on machine learning were developed to predict surface quality from photographs of Ti6Al4V parts created by single point incremental forming. Training data were collected using a surface profilometer and a stereoscopic microscope, and more than one hundred and fifty simulations were run to optimize the processing parameters and to find the most suitable classifier. The main findings derived from the current work are as follows:

- The support vector machine has proven to be a competent classifier to predict the surface finish according to the methodology used.
- The best results are achieved using second- and third-degree polynomial kernels.
- Working with grayscale images emphasizes the most advisable features to measure the surface roughness.
- Dividing photographs to increase the database does not justify the deviation between images and actual values when average parameters are measured.

Overcoming the limitations of this work will provide the basis for the use of a regression model, instead of a classification one, to predict continuous values. One future direction will be to build a large and uniform database to avoid a class imbalance and to apply normalization and filtering techniques in order to obtain stable roughness values and to reduce the noise in pictures, respectively. In this way, it will be possible to build a reliable model with high detection capacity to be deployed in a productive environment.

In the future, this measuring strategy is expected to be used on real-time video transmission in a manner that enables automatic and fast selection of parts according to tolerance requirements.

Author Contributions: V.M. and F.G.-S.: Conceptualization. V.M., F.G.-S., F.B.-M. and J.N.: Data curation. V.M., F.G.-S. and F.B.-M.: Formal analysis. F.G.-S. and V.M.: Funding acquisition. V.M., F.G.-S. and J.N.: Investigation. F.G.-S., V.M. and F.B: Methodology. V.M.: Project administration. V.M.: Resources. F.G.-S. and F.B.-M.: Software. V.M., F.G.-S. and M.C.M.: Supervision. V.M. and F.G.-S.: Validation. F.B.-M., F.G.-S. and V.M.: Visualization. V.M., F.G.-S., F.B.-M., J.N. and M.C.M.: Writing—original draft. V.M., F.G.-S. and M.C.M.: Writing—review and editing. All authors have read and agreed to the published version of the manuscript.

Funding: This research received no external funding.

Institutional Review Board Statement: Not applicable.

Informed Consent Statement: Not applicable.

Conflicts of Interest: The authors declare no conflict of interest.

References

1. Silva, M.; Skjoedt, M.; Bay, N.; Martins, P.A.F. Revisiting single-point incremental forming and formability/failure diagrams by means of finite elements and experimenta-tion. *J. Strain. Anal. Eng. Des.* **2019**, *44*, 221–234. [CrossRef]
2. Araújo, P.; Teixeira, L.; Montanari, A.; Reis, A.; Silva, M.B.; Martins, P.A.F. Single point incremental forming of a facial implant. *Prosthet. Orthot. Int.* **2014**, *38*, 369–378. [CrossRef]
3. Vanhove, H.; Carette, Y.; Vancleef, S.; Duflou, J.R. Production of thin shell clavicle implants through single point incremental forming. *Procedia Eng.* **2017**, *183*, 174–179. [CrossRef]
4. Ambrogio, G.; De Napoli, L.; Filice, L.; Gagliardi, F.; Muzzupappa, M. Application of incremental forming process for high customised medical product manufacturing. *Mater. Process. Technol.* **2005**, *162–163*, 156–162. [CrossRef]
5. Saidi, B.; Giraud-Moreau, L.; Cherouat, A.; Nasri, R. Experimental and numerical study on optimization of the single point incremental forming of AISI 304L stainless steel sheet. *J. Phys. Conf. Ser.* **2017**, *896*, 012039. [CrossRef]
6. Götmann, A.; Bailly, D.; Bergweiler, G.; Bambach, M.; Stollenwerk, J.; Hirt, G.; Loosen, P. A novel approach for temperature control in ISF supported by laser and resistance and resistance heating. *Int. J. Adv. Manuf. Technol.* **2013**, *67*, 2195–2205. [CrossRef]
7. Duflou, J.R.; Callebaut, B.; Verbert, J.; De Baerdemaeker, H. Improved SPIF performance through dynamic local heating. *Int. J. Mach. Tools* **2008**, *48*, 543–549. [CrossRef]
8. Fan, G.; Sun, F.; Meng, X.; Gao, L.; Tong, G. Electric hot incremental forming of Ti6Al4V titanium sheet. *Int. J. Adv. Manuf. Technol.* **2010**, *49*, 941–947. [CrossRef]
9. Honarpisheh, M.; Abdolhoseini, M.J.; Amini, S. Experimental and numerical investigation of the hot incremental forming of Ti6Al4V using electrical current. *Int. J. Avd. Manuf. Technol.* **2016**, *83*, 2027–2037. [CrossRef]
10. Ambrogio, G.; Filice, L.; Gagliardi, F. Formability of lightweight alloys by hot incremental sheet forming. *Mater. Des.* **2012**, *34*, 501–508. [CrossRef]
11. Naranjo, J.; Miguel, V.; Martínez-Martínez, A.; Coello, J.; Manjabacas, M.C. Evaluation of the Formability and Dimensional Accuracy Improvement of Ti6A14V in Warm SPIF Processes. *Metals* **2019**, *9*, 272. [CrossRef]

12. Ortiz, M.; Penalva, M.; Iriondo, E.; López de Lacalle, L. Accuracy and Surface Quality Improvements in the Manufacturing of Ti-6Al-4V Parts Using Hot Single Point Incremental Forming. *Metals* **2019**, *9*, 697. [CrossRef]
13. Durante, M.; Formisano, A.; Langella, A.; Memola Capece Minutolo, F. The influence of tool rotation on an incremental forming process. *J. Mat. Process. Technol.* **2009**, *209*, 4621–4626. [CrossRef]
14. Palumbo, G.; Brandizzi, M. Experimental investigations on the single point incremental forming of a titanium alloy component combining static heating with high tool rotation speed. *Mater. Des.* **2012**, *40*, 43–51. [CrossRef]
15. Lu, B.; Fang, Y.; Xu, D.; Chen, J.; Ou, H.; Moser, N.; Cao, J. Mechanism investigation of friction-related effects in single point incremental forming using a developed oblique roller-ball tool. *Int. J. Mach. Tools Manuf.* **2014**, *85*, 14–29. [CrossRef]
16. Jawale, K.; Duarte, J.; Reis, A.; Silva, M. Microstructural investigation and lubrication study for single point incremental forming of copper. *Int. J. Solids Struct.* **2018**, *151*, 145–151. [CrossRef]
17. Moreas, G.; van de Velde, F.; Bilstein, W. Advanced sensor for on-line topography in continuous annealing lines. *Rev. Metall. Cah. Inf. Techn.* **2006**, *103*, 233–237. [CrossRef]
18. Lehmann, P. Surface-roughness measurement based on the intensity correlation function of scattered light under speckle-pattern illumination. *Appl. Opt.* **1999**, *38*, 1144–1152. [CrossRef] [PubMed]
19. Abd Ali, R.; Chen, W.; Al-Furjan, M.; Jin, X.; Wang, Z. Experimental investigation and optimal prediction of maximum forming angle and surface roughness of an Al/SUS bimetal sheet in an incremental forming process using machine learning. *Materials* **2019**, *12*, 4150. [CrossRef]
20. Echrif, S.; Hrairi, M. Significant Parameters for the Surface Roughness in Incre-mental Forming Process. *Mater. Manuf. Process.* **2014**, *29*, 697–703. [CrossRef]
21. Mulay, A.; Ben, B.; Ismail, S.; Kocanda, A. Prediction of average surface rough-ness and formability in single point incremental forming using artificial neural network. *Arch. Civ. Mech. Eng.* **2019**, *19*, 1135–1149. [CrossRef]
22. Abu-Mahfouz, I.; El Ariss, O.; Esfakur Rahman, A.; Banerjee, A. Surface roughness prediction as a classification problem using support vector machine. *Int. J. Adv. Manuf. Technol.* **2017**, *92*, 803–815. [CrossRef]
23. Koblar, V.; Pecar, M.; Gantar, K.; Tusar, T.; Filipic, B. Determining Surface Roughness of Semifinished Products Using Computer Vision and Machine Learning. In *Proceedings of the 18th International Multiconference Information Society*; Information Society: Ljubljana, Eslovenia, 2015; pp. 51–54.
24. Kurra, S.; Rahman, N.H.; Regalla, S.P.; Gupta, A.K. Modeling and optimization of surface roughness in single point incremental forming process. *J. Mater. Res. Technol.* **2015**, *4*, 304–313. [CrossRef]
25. Lin, W.; Lo, S.; Young, H.; Hung, C. Evaluation of Deep Learning Neural Networks for Surface Roughness Prediction Using Vibration Signal Analysis. *Appl. Sci.* **2019**, *9*, 1462. [CrossRef]
26. Li, Z.; Zhang, Z.; Shi, J.; Wu, D. Prediction of surface roughness in extrusion-based additive manufacturing with machine learning. *Robot. Comput. Integr. Manuf.* **2019**, *57*, 488–495. [CrossRef]
27. Simeone, O. A Brief Introduction to Machine Learning for Engineers, Found. *Trends Signal. Process.* **2018**, *12*, 200–431. [CrossRef]
28. Badillo, S.; Banfai, B.; Birzele, F.; Davydov, I.; Hutchinson, L.; Kam-Thong, T.; Zhang, J. An introduction to machine learning. *Clin. Pharmacol. Ther.* **2020**, *107*, 871–885. [CrossRef]
29. Utkin, L.V. An imprecise extension of SVM-based machine learning models. *Neurocomputing* **2019**, *331*, 18–32. [CrossRef]
30. Mathworks®. Available online: www.mathworks.com/help/stats/choose-a-classifier.html (accessed on 6 June 2021).
31. Moreas, G.; Bilstein, W. On-line industrial roughness and topography measurement for continuous lines. In *CCATM'2010, The Fifteenth CSM Conference and Exhibition on Analysis and Testing of Materials*; CSM: Beijing, China, 2010.

Article

The Performance of 3D Printed Polymer Tools in Sheet Metal Forming

Fabio Tondini, Alberto Basso , Ulfar Arinbjarnar and Chris Valentin Nielsen *

Department of Mechanical Engineering, Technical University of Denmark, 2800 Kongens Lyngby, Denmark;
s192925@student.dtu.dk (F.T.); albass@mek.dtu.dk (A.B.); ulari@mek.dtu.dk (U.A.)
* Correspondence: cvni@mek.dtu.dk

Abstract: Additively manufactured polymer tools are evaluated for use in metal forming as prototype tools and in the attempt to make sheet metal more attractive to small production volumes. Printing materials, strategies and accuracies are presented before the tools and tested in V-bending and groove pressing of 1 mm aluminum sheets. The V-bending shows that the tools change surface topography during forming until a steady state is reached at around five strokes. The geometrical accuracy obtained in V-bending is evaluated by the spring-back angle and the resulting bend radius, while bending to 90° with three different punch nose radii. The spring-back shows additional effects from the elastic deflection of the tools, and the influence from the punch nose radius is found to be influenced by the printing strategy due to the ratio between tool radius and the printed solid shell thickness enclosing the otherwise less dense bulk part of the tool. Groove pressing shows the combined effect of groove heights and angular changes due to spring-back. In all cases, the repeatability is discussed to show the potential of tool corrections for obtaining formed parts closer to nominal values.

Keywords: additive manufacturing; rapid prototyping; sheet metal forming; V-bending; groove pressing

Citation: Tondini, F.; Basso, A.; Arinbjarnar, U.; Nielsen, C.V. The Performance of 3D Printed Polymer Tools in Sheet Metal Forming. *Metals* **2021**, *11*, 1256. https://doi.org/10.3390/met11081256

Academic Editor: Golden Kumar

Received: 30 June 2021
Accepted: 7 August 2021
Published: 9 August 2021

Publisher's Note: MDPI stays neutral with regard to jurisdictional claims in published maps and institutional affiliations.

1. Introduction

Sheet metal forming can be used to create a variety of complicated three-dimensional parts from flat sheet metal by, e.g., bending and deep drawing. Some of the advantages of sheet metal forming are the beneficial mechanical properties of the formed part, the minimal amount of scrap that is generated and the high production capacity that can be attained. The main disadvantage of the method is the cost and time necessary for setting up a new process line. To ensure economic production, sheet metal forming is therefore often limited to mass production [1].

One of the major costs related to sheet metal forming is the development and production of the forming tools. Conventionally, the tool is designed by a tool designer and machined out of metal, which can take a long time and be quite costly. During tool development, a tool may then go through one or more revisions before reaching its final form. Each revision would necessitate a new round of machining or other processing steps each time. This quickly adds up to massive costs, both money and time-wise.

Additive manufacturing has gained popularity as a method of rapid prototyping (RP) tool concepts. These prototypes can then serve as either a proof-of-concept or as tools for small series production [2]. Adopting this method reduces the time and money spent on the tool, and allows for increased flexibility in sheet metal forming [1]. Zaragosa et al. [3] found that implementing an RP solution is more expensive than fabricating a single metal tool due to the initial costs of purchasing equipment and configuring it. However, the cost of printing further tools becomes much smaller than the cost of machining metal tools. It is unlikely that a first tool design is perfect due to the increasing complexity of parts and new sheet materials, which, as noted by Durgun [4], makes the experience of conventional tool makers insufficient. Therefore, inevitable revisions and iterations on the tool design become cheaper with 3D printing of tools.

Various researchers have investigated the suitability of using 3D-printed tools for sheet forming. Zaragosa et al. [3] studied the use of 3D-printed tool inserts in a metal frame for use in V-bending and found the spring-back to be similar to using fully metal tools. Nakamura et al. [5] used fully plastic tools for V-bending and found that the accuracy of the final bend angle is reduced compared to metal tools due to the increased elastic deflection, nevertheless, the spring-back was repeatable for each material. They further noted that the spring-back remained mostly constant for the plastic tools over 100 bending operations. Nakamura et al. [5] also found, and Zaragosa et al. [3] later confirmed, that using plastic tools results in a better surface finish on the formed part compared to using metal tools. This is because the workpiece is unlikely to be scratched by the plastic tools. Klimyuk et al. [1] optimized their printing strategy and managed to deep draw a cup, noting that no wear was visible on the surface of the printed tools. Schuh et al. [6], on the other hand, after optimization of the printing strategy using a Design of Experiments approach, managed to deep draw a cup but noted that most of the wear occurs during the first operation. Aksenov and Kononov [7] fabricated heat exchanger plates using printed tools and noted that lubrication was not necessary, as the plastic had anti-friction properties. In all the above references, there is agreement that 3D printing of forming tools is suitable for rapid prototyping and small series production.

V-bending is one of the simplest sheet forming operations. It involves a V-shaped punch that presses a sheet into a die, forcing the sheet to bend to a predefined angle. The V-shape is typically designed to account for spring-back, and so that the radius of the corner of the bend is within the tolerance of part design. The ability to bend the sheet close to the nominal angle and dimension depends on the tool, meaning that the spring-back that occurs and the accuracy of the formed radius can be used as an indicator of tool performance [3].

In this work, the performance of 3D-printed tools is investigated under sheet metal forming conditions. After optimizing the printing strategy through V-bending tests and surface characterization, the optimized printing strategy is used to fabricate tools for a groove pressing process. This process is used to form a geometry that is more complex than the basic V-bending geometry to demonstrate the types of geometries that it is possible to form using polymer tools. Further, it serves as another way of evaluating the accuracy of formed parts when using polymer tools, bringing this method closer to an industrial application.

2. Materials and Methods

2.1. Equipment and Evaluation Geometries

The sheet material used in all parts of this study is commercially pure aluminum, EN AW-1050A H14, of 1.0 mm thickness (Sanistål, Aalborg, Denmark). Two forming processes are considered, V-bending and groove pressing. The initial blanks for each process have dimensions of 30 mm $\times$ 70 mm and 30 mm $\times$ 90 mm, respectively, with the target geometries of the processes shown in Figure 1.

Both forming processes were carried out in an electric press with a maximum capacity of 150 kN. A sub-press, designed for use with interchangeable tool inserts and is shown in Figure 2, is used to ensure proper guidance of the printed forming tools during forming. All experiments were conducted to fixed end positions with a vertical press speed of 3 mm/s.

All forming tools used in this study were printed using a Fused Filament Fabrication (FFF) printer of the Ultimaker 2+ type (Ultimaker, Utrecht, The Netherlands). This printer extrudes a polylactic acid (PLA) filament of $\varnothing 2.85$ mm through an $\varnothing 0.4$ mm nozzle. From new, it has an XY-positioning (plane) accuracy of 12.5 µm and a Z-positioning (vertical) accuracy of 5 µm [8]. Using Vat Photopolymerization Additive Manufacturing (VPAM) as a method of producing tools was also investigated. The principle of the method is that liquid resin is selectively cured by a UV light in a layer-by-layer fashion. The printer used in this case was an Elegoo Mars 2 Pro, which has an XY-positioning (plane) accuracy of 50 μm, and an accuracy of 1 μm along the Z-axis (vertical).

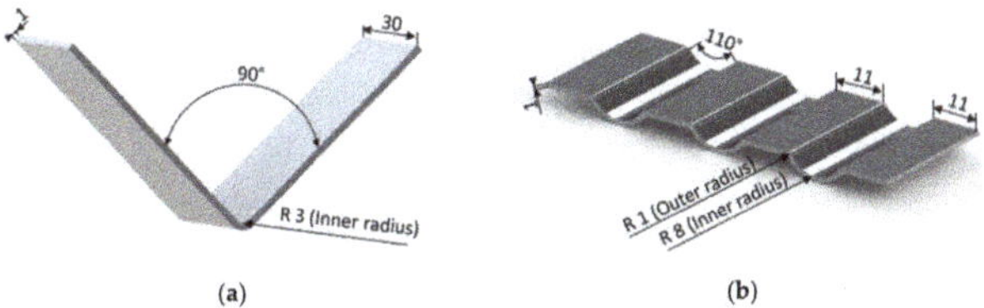

Figure 1. Target geometries with dimensions in mm: (**a**) sheet formed by V-bending; (**b**) sheet formed by groove pressing.

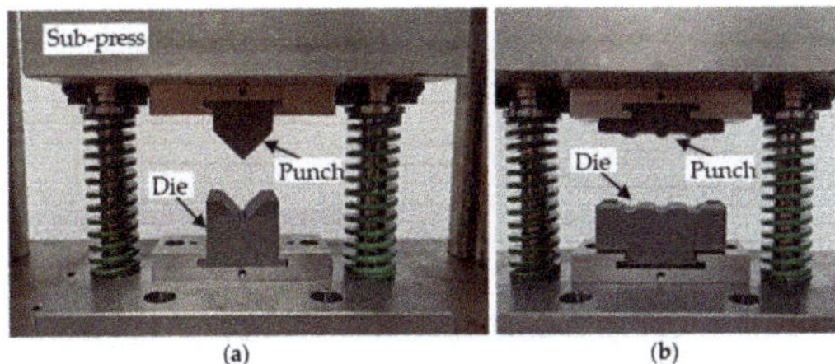

Figure 2. Sub-press with tools mounted for (**a**) V-bending and (**b**) groove pressing.

2.2. Printing Material and Strategy

Among the different additive manufacturing processes, Fused Filament Fabrication (FFF) and Vat Photopolymerization (VPAM) are widely used in the field of rapid prototyping. In this work, both printing methods are investigated. FFF is an additive manufacturing process, where a molten polymer filament is extruded through a nozzle onto a build platform, layer by layer. In VPAM, a liquid resin is selectively cured through light activated polymerization by use of a UV light. Low costs and high manufacturing speeds are the core of rapid prototyping, so for the purpose of this study, the most commonly available materials were selected for both FFF and VPAM. PLA was used for the tools produced by FFF; this material is the most commonly used in FFF, it is biodegradable and produced from renewable resources such as corn starch and sugar cane. Furthermore, it is characterized by relatively low melting temperature and low shrinkage, reducing internal residual stresses [9]. The resin selected for VPAM is an acrylate-based photopolymer with phosphine oxide photo initiator. This photopolymer is affordable and is characterized by a great trade-off between market price and mechanical and printing properties. It was selected for its high hardness after curing compared to similar resins: 83 shore D hardness according to its datasheet. Some mechanical and physical properties of the materials that that are used in this work are shown in Table 1.

Table 1. Typical mechanical and physical properties of materials that are used in this work.

Material	ρ (kg/m^3)	E (GPa)	ν (-)
AL1050	2710	71	0.3
Polylactic acid (PLA)	1230	2	0.3
Photopolymer	1120	2.7	0.36

The final strength and quality of the 3D-printed tools are directly influenced by various printing parameters [1,10–13]. Klimyuk et al. [1] showed how layer thickness, wall thickness and infill density have a major impact on the final quality of the printed part. The authors of this work conducted an investigation on the influence of printing orientation, layer thickness, wall thickness and infill density in order to select the optimal parameters for printing the tools. In the case of VPAM, the effect of the exposure time was also studied. The selected printing parameters are shown in Table 2.

The printing orientation is an important parameter due to the anisotropy of the properties of the printed part, which is generated by its layer structure [9]. The printing orientation was chosen to be in the out of plane direction in Figure 2, because parts printed in this direction withstand compressive forces during metal forming better by having the layers perpendicular to vertical forces. This minimizes the risk of delamination between layers.

The selected layer thickness is 0.1 mm. In FFF, thinner layers lead to improved adhesion of layers to one another and denser parts since the heat from the nozzle, being closer to the previous layer, helps the layers bond together. Kuznetsov et al. [9] showed that by decreasing the layer thickness, it is possible to decrease the generation of voids, which negatively affect layer adhesion and, therefore, tool strength. The layer thickness is also an important parameter for VPAM, where a smaller layer thickness ensures complete UV penetration through the layer and a higher degree of polymerization [11].

The shell thickness was fixed to 2.10 mm (double of the default wall thickness of 1.05 mm). Aslani et al. [12] investigated the influence of the shell thickness on the geometrical accuracy of the part showing higher accuracy when a double wall thickness was used.

The infill density is another important parameter. The higher the infill density, the larger the amount of material that the component is comprised of, and the greater the bonding areas between layers. The larger bonding area leads to an adhesion gain among the layers, reducing the likelihood of delamination and fracture and promoting the strength of the component. An infill density of 50% was selected for FFF as a balance between printing time/material volume and component strength. The infill density was set to 100% for the VPAM tools in order to ensure complete polymerization and to maximize the bonding between layers, reducing the risk of delamination.

The exposure time in VPAM directly affects the light penetration depth, which is the length by which the UV light is able to penetrate and cure the photopolymer. The light penetration depth needs to be equal to, or more than, the layer thickness in order to assure bonding between the layers. The optimal exposure time, for the specific resin used in this work, was found to be 30 s for the burn in layers and 3.5 s for the rest of the layers.

Table 2. Printing process parameters for FFF and VPAM.

Printing Technology	Exposure Time	Orientation	Layer Thickness	Shell Thickness	Infill Density	Post Processing
FFF	-	Vertical	0.1 mm	2.10 mm	50%	-
VPAM	3.5 s	Vertical	0.1 mm	-	100%	UV curing at 45 °C for 30 min

2.3. Precision, Accuracy and Surface Roughness of the Printed Parts

A study on the accuracy of the two printers was carried out after the printing parameters had been selected. The reference geometry in Figure 3 was printed five times with FFF and five times with VPAM and measured to evaluate accuracy and repeatability. Figure 4 shows the deviation in x, y and z directions from the nominal dimensions. The largest deviations can be detected in x and y directions on the parts printed with VPAM, showing how the photopolymer is more sensitive to shrinkage in comparison to PLA.

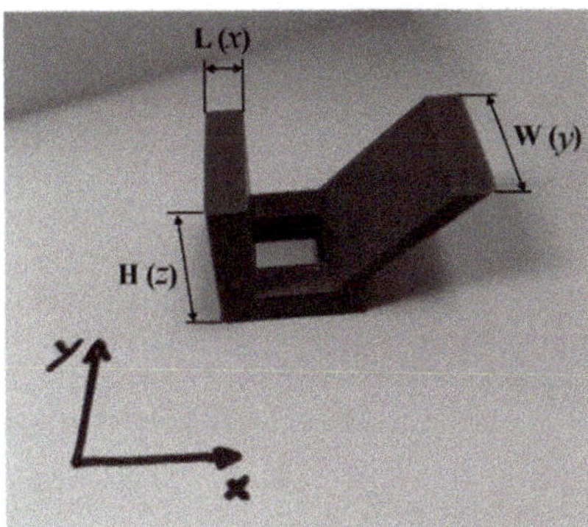

Figure 3. Reference geometry used to determine printer accuracy with dimensions L × H × W: 5 mm × 20 mm × 20 mm. The printing direction is in *z*.

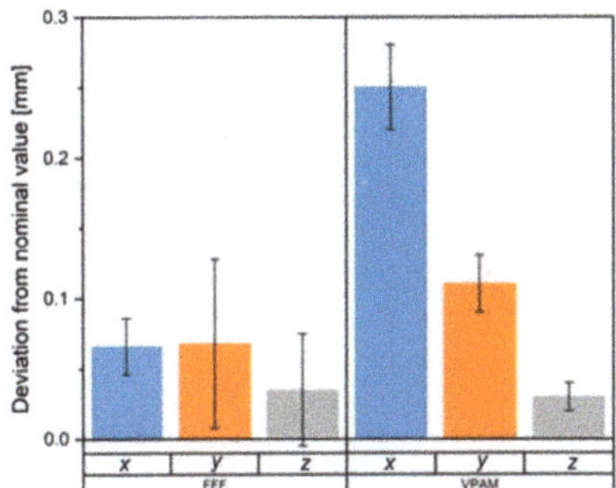

Figure 4. Deviation in *x*, *y* and *z* directions of reference geometries printed with FFF and VPAM.

The standard deviation bars in Figure 4 depict the accuracy of the two printers relative to dimensions given in Figure 3. Overall, the VPAM printer showed better accuracies in comparison to FFF Accuracies of 0.03 mm and 0.02 mm were detected, respectively, for the x and y directions with the VPAM printer, while FFF showed accuracies of 0.02 mm in *x* and 0.06 mm in *y*. A particularly good accuracy was detected in the z direction for the VPAM with a value of 0.01 mm.

The surface roughness of the printed tools was investigated by using an Olympus LEXT confocal microscope. The linear mean height *Ra* of the tools was evaluated parallel and perpendicular to the printing direction. The *Ra* is comparable for both FFF and VPAM when looking at the surface parallel to the printing direction, with a value of 0.78 ± 0.45 μm and 0.81 ± 0.25 μm, respectively. The investigation of the roughness on the surface perpendicular to the printing direction showed instead higher *Ra* for FFF compared to VPAM, having *Ra* values of 2.96 ± 0.27 μm and 1.43 ± 0.21 μm, respectively. Even though VPAM was found to be more accurate than FFF, this additive manufacturing technology needs post-processing after the printing, increasing time and cost for the production of the tool. Therefore, FFF was selected as the preferred method of RP for further investigation.

2.4. Measurement Procedures and Strategy

The objective of this work is to study the geometrical accuracy of parts formed by RP polymer tools. The final shape of formed components and RP tools is compared to the nominal shape by use of a ZEISS DuraMax coordinate measuring machine. All measurements that are reported are averaged along the entire width of the respective component.

The nominal geometry of the component created by V-bending is shown in Figure 1a. The measurement strategy used to determine the geometrical accuracy of the formed component is outlined in Figure 5a. Five measurements of angles (M1–M5) are performed along the arms of the bent components at equal intervals of 6 mm. The radius (R) where the arms meet is also measured.

The geometrical accuracy of the part formed by groove pressing, shown in Figure 1b, is determined by the strategy outlined in Figure 5b. The angularity of the four flat parts (P1–P4) is measured with respect to the outer tangential plane touching the planes P2 and P3. The height of the three main grooves (H1, H2, H3) were measured with respect to the outer tangential plane touching the planes P1 and P2, P2 and P3, P3 and P4, respectively. The overall angularity of the workpiece is measured between P1 and P4.

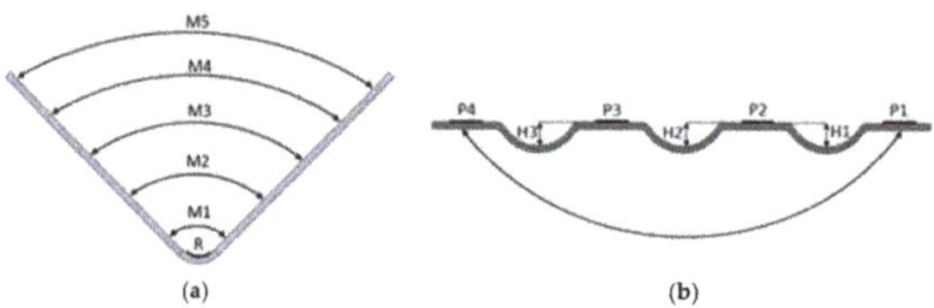

Figure 5. Measurement strategies for sheets formed by (**a**) V-bending and (**b**) groove pressing.

For evaluating the geometrical accuracy of the printed punch, as shown in Figure 6, the strategy consists of five angle measurements (M1–M5) located at an equal distance from each other of 5 mm on the tilted surfaces and in five radius measurements (R1–R5) on the punch nose located at an equal distance of 8 mm from each other across the width.

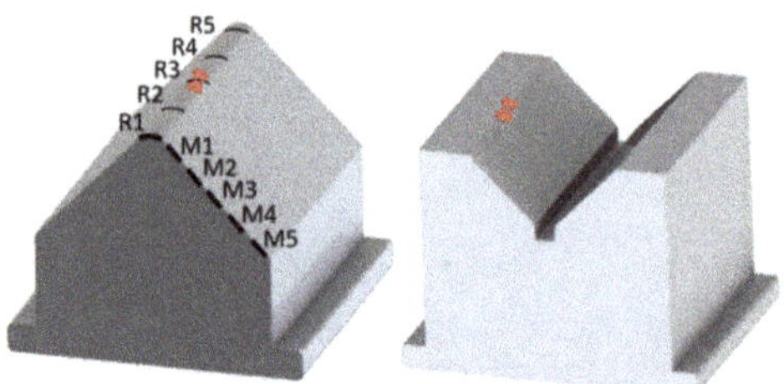

Figure 6. V-bending punch measurement locations and surface inspection locations (red marks) on both punch and die.

3. Accuracy and Wear in V-Bending

3.1. Change in Tool Geometry and Surface vs. Number of Strokes

A visual inspection at selected points of the tool surfaces was performed using a LEXT OLS4000 confocal microscope (Olympus, Tokyo, Japan). The locations, indicated by the red marks in Figure 6, were selected as the highest tribological load would be found there. The resulting pictures are shown in Table 3, as a function of the number of bending strokes. The picture taken for stroke #0 is not necessarily in the same location as others, as there were no existing marks which could be used as a reference point. It does, however, show that the surface is not perfect even before forming, with defects possibly having been caused by improper handling. Pictures taken for strokes #3–30 are taken in the same location to showcase how the wear progresses. The figures show that the punch does not experience significant changes to the surface, while the die does. The surface of the die changes more gradually after the fifth stroke, which may indicate some form of steady state. There is considerably less sliding occurring in the interface between the punch and the workpiece than in the interface between the die and the workpiece, which is likely the reason for the difference in wear between the two tools.

Table 3. Images of tool surfaces for different numbers of strokes that have been performed. Each figure shows an area that is 600 μm by 600 μm.

Stroke (#)	0	3	5	10	30
Punch					
Die					

3.2. Springback, Geometrical Accuracy and Comparison to Tool Accuracy

The results of the geometrical accuracy study of the workpiece and punch, obtained by employing the measurement strategy described in Section 2.4, are shown in Figures 7 and 8.

Figure 7 shows the geometrical accuracy of the 3D-printed punches after they have been used to form five parts. The mean angle that is measured is $90.2 \pm 0.1°$. The figure also shows the averaged angle over measurements of parts formed in five consecutive strokes. The angle is smaller than nominal and grows with distance from the punch nose. The deviation from nominal is largest at the punch nose. This behavior is most obvious for the tool with a 1 mm radius, where the mean angle grows from $86.1°$ to $89.8°$ as the measurements are taken further away from the punch nose.

Figure 8 shows the average angle measured in location M3 of the workpieces, over five strokes, for the three different punch nose radii. The largest difference from nominal is found when using the 3 mm radius tool after three strokes, where the measured value is $88.2 \pm 0.8°$. The smallest deviation from nominal is found when using the 1 mm radius tool at the second stroke, where the measured value is $88.7 \pm 0.27°$. The workpieces formed by the 2 mm radius tool have small deviations among them and a steady response over the five strokes, characterized by a standard deviation of $0.08°$. This behavior is not observed for the workpieces formed by the 1 mm and 3 mm radii tools.

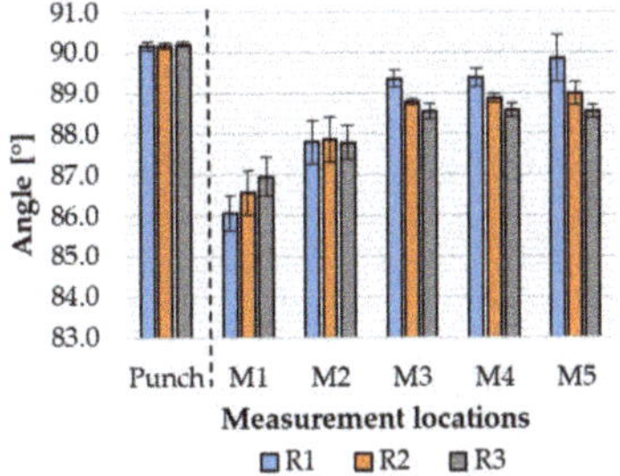

Figure 7. Measured angles as function of measurement location for tools with different nose radii.

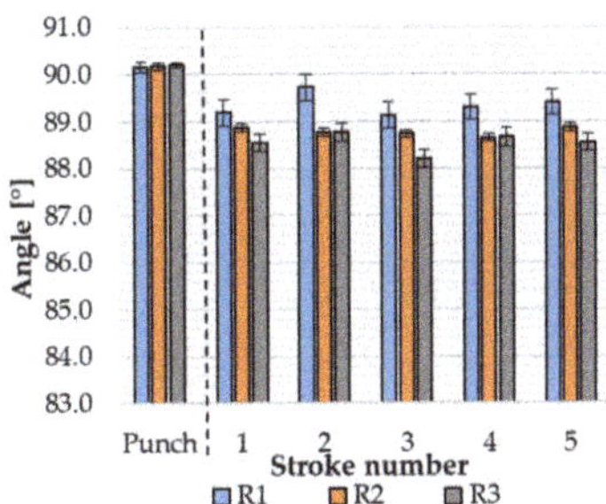

Figure 8. Measured angle at location *M3* as function of no. strokes performed for tools with different nose radii.

3.3. Bend Radius

Another aspect of the accuracy is the bend radius that is achievable by 3D-printed tools. Three different radii are tested in V-bending, with results shown in Figures 9 and 10. The bend radii of the workpieces as function of the number of strokes in the same tool is shown in Figure 9. Figure 10 shows a comparison between the radius of the printed tools and the radius measured on the formed components. The nose radii of the R1 and R2 tools are not nominal, but the formed component adopts the measured nose radius closely. The scatter of measurements is also low, indicating that the repeatability is good. This implies that the deviation from nominal in the formed components is due to the accuracy of the printed tools, which could be compensated for. The R3 tool is different, in that the radius of the tool is close to nominal, but the radius of the formed component is not. The scatter in the measured radius is also larger than for components formed using the R1 and R2 tools. The shell thickness that is used when printing the tools is 2.1 mm for all punches regardless of the punch nose radius. As larger punch nose radius implies a larger contact area towards the workpiece, the constant shell thickness results in less stiff behavior from the punch nose when the radius is increased. The corresponding larger elastic deflection of the R3 tool would explain the increased radius and scatter measured on the bent workpieces.

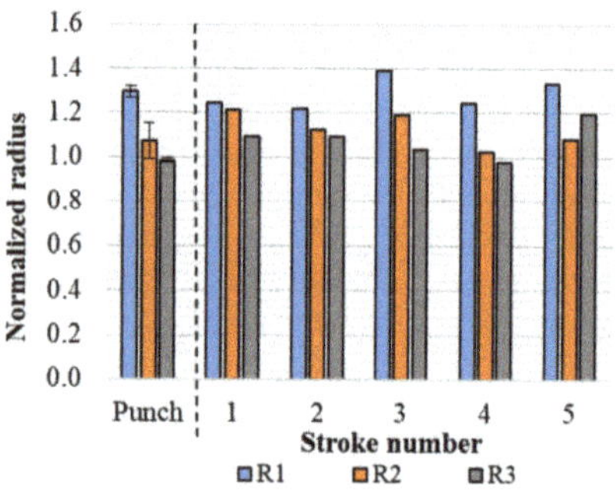

Figure 9. Measured radius of workpiece as function of no. strokes and of punches normalized with respect to nominal dimensions.

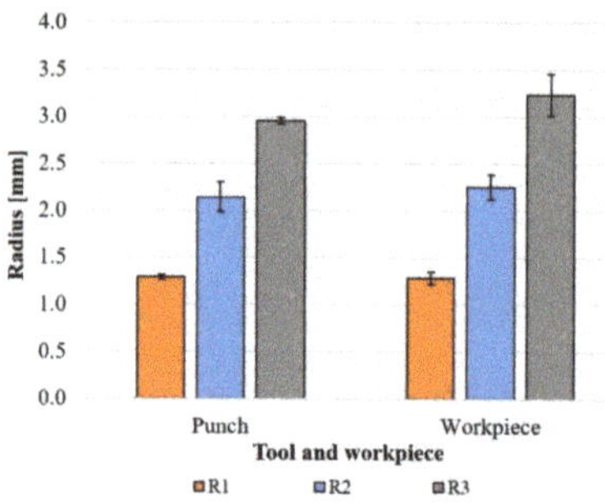

Figure 10. Average radius of workpiece over five strokes and of punches with different nose radii.

4. Groove Pressing

The geometrical accuracy obtained by groove pressing (Figure 1b) is evaluated by the measurement strategy shown in Figure 5b. The geometrical accuracy of the tools, shown in Figure 2b, is also evaluated so that the analysis can be corrected for tool dimensions deviating from nominal. Two key features were selected for evaluating the workpiece accuracy: the angularity of the four flat parts and the height of the three grooves.

Figure 11 shows that the tools are printed with slight increasing height from H1 to H3. This is probably due to a slight tilting of the building plate during printing. The measured heights of the workpiece groove heights were relatively close to the heights imposed by the tools. The workpiece groove heights are very repeatable, with standard deviations between 0.6–1.1% across the three groves.

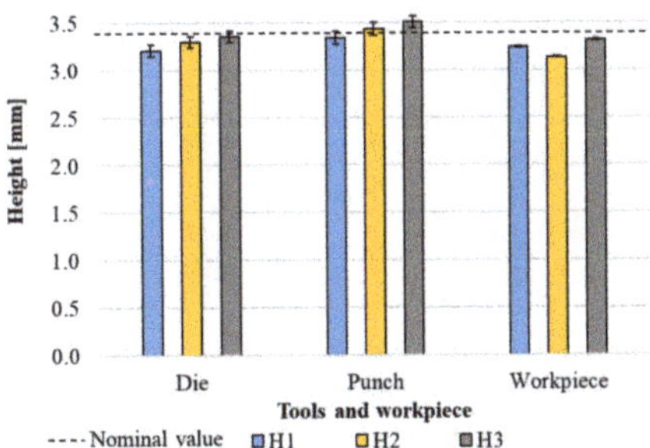

Figure 11. Average groove height in workpiece and tools compared to the nominal height 3.4 mm.

Figure 12 shows the angularity measured between the features that nominally sit on the same plane. The nominal angularity is therefore 180°. The figure shows that the tools are close to nominal. The average angularity measured on the workpiece is far from the nominal value of 180° due to spring-back and varies between the features. The largest deviation was observed on P4, which has an average value of $178.3 \pm 0.12°$. P3 has the closest value to nominal by $179.2 \pm 0.11°$. Spring-back after the nominally symmetric groove pressing is expected to increase angularity deviation from nominal with distance from the center. This explains why the deviation is larger in P4 than in P3. However, Figure 12 also shows that this cannot explain P1 and P2, because there is no symmetry in the resulting angularities. Since the tools are symmetrical in terms of angularity with respect to each other, the asymmetric angularity in the workpiece is due to the asymmetric groove heights in the tools, as shown in Figure 11.

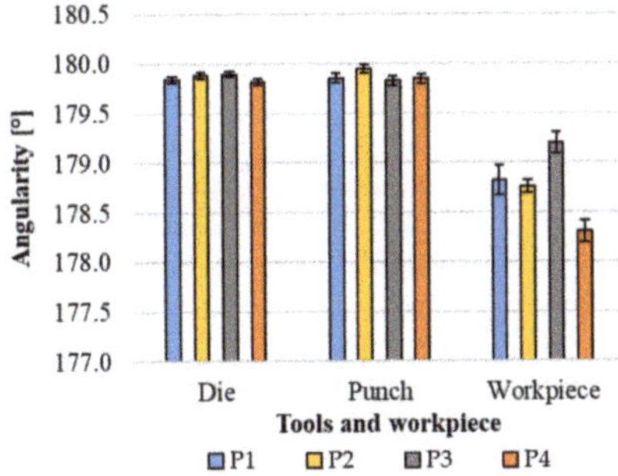

Figure 12. Average angularity of workpiece and tools for the four key planes.

The performance of the tools was also investigated over a small number of strokes. In this case, as mentioned in Section 2.4, the overall workpiece angularity is described by the angle between P1 and P4 and the reported height is an average value of H1, H2 and H3. Figure 13 shows the measured dimensions normalized with respect to the nominal values. The stability and repeatability of the measured values are good.

Regarding the height measurements, it is noticed that the standard deviations of both tools and workpieces are around 4%. The angularity is more stable and has standard deviations of 0.2%. The normalization shows that the angularity of the entire part is much closer to the nominal value than the height of the grooves. This is clearly linked to the deviation in the die, which could be compensated for. The angularity of the tools could also be altered by a small curvature to result in angularity of the workpieces after spring-back even closer to nominal.

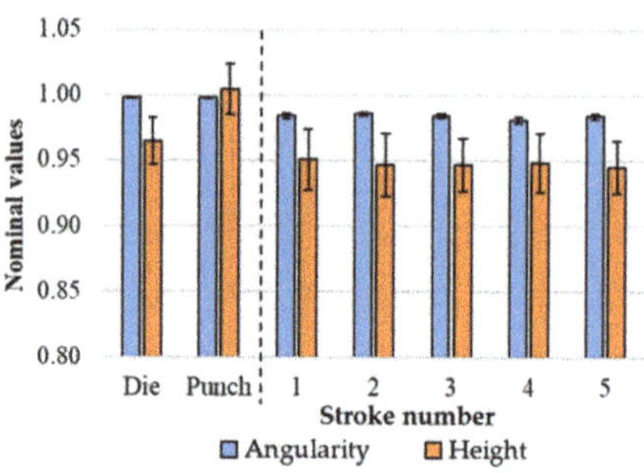

Figure 13. Normalized measurements of workpiece and tool feature dimensions.

5. Conclusions

This paper demonstrates the application of FFF to produce tools for sheet metal forming, exemplified by V-bending and grove pressing of aluminum sheets. The following conclusions can be drawn:

- Steady-state conditions were achieved in five strokes for the tool assembly with minimal wear detected in following strokes.
- The elastic deflection of the tools influence forming stroke and the final geometry after spring-back.
- The radius of the punch nose in V-bending should be considered relative to the shell-thickness used in printing the tools for optimal results.
- It is possible to form more complex shapes, which was here exemplified by groove pressing.
- Some compensation for spring-back will be possible due to low scatter in the final geometries.

The results of this work show the high potential of polymer tools in sheet metal forming for small series or prototyping, but also that there is room for improvement and further research in the field.

Author Contributions: Conceptualization, F.T., A.B., U.A. and C.V.N.; methodology, F.T., A.B. and C.V.N.; software, U.A.; formal analysis, F.T.; investigation, F.T., A.B., U.A. and C.V.N.; data curation, F.T., A.B. and U.A.; writing—original draft preparation, F.T., A.B. and U.A.; writing—review and editing, F.T., A.B., U.A. and C.V.N.; visualization, F.T. and U.A.; supervision, A.B., U.A. and C.V.N.; project administration, C.V.N.; funding acquisition, C.V.N. All authors have read and agreed to the published version of the manuscript.

Funding: This research was partly funded by the Danish Council for Independent Research, grant number DFF—0136-00159A. The APC was funded by the Technical University of Denmark.

References

1. Klimyuk, D.; Serezhkin, M.; Plokikh, A. Application of 3D printing in sheet metal forming. *Mater. Today Proc.* **2021**, *38*, 1579–1583. [CrossRef]
2. Zelený, P.; Vána, T.; Stryal, J. Application of 3D printing for specific tools. *Mater Sci Forum.* **2016**, *862*, 316–323. [CrossRef]
3. Zaragosa, V.G.; Strano, M.; Iorio, L.; Monno, M. Sheet metal bending with flexible tools. *Proc. Manuf.* **2019**, *29*, 232–239. [CrossRef]
4. Durgun, I. Sheet metal forming using FDM rapid prototype tool. *Rapid Prototyp. J.* **2015**, *21*, 412–422. [CrossRef]
5. Nakamura, N.; Mori, K.; Abe, F.; Abe, Y. Bending of sheet metals using plastic tools made with 3D printer. *Proc. Manuf.* **2018**, *15*, 737–742. [CrossRef]

6. Schuh, G.; Bergweiler, G.; Bickendorf, P.; Fiedler, F.; Colag, C. Sheet metal forming using additively manufactured polymer tools. *Proc. CIRP* **2020**, *93*, 20–25. [CrossRef]
7. Aksenov, L.B.; Kononov, I.Y. Thin sheet forming with 3D printed plastic tool. *Solid State Phenom.* **2020**, *299*, 705–710. [CrossRef]
8. Dynamism, Ultimaker 2+. Available online: https://www.dynamism.com/ultimaker/ultimaker-2-plus.html (accessed on 29 April 2021).
9. Kuznetsov, V.E.; Solonin, A.N.; Urzhumtsev, O.D.; Shilling, R.; Tavitov, A.G. Strength of PLA components fabricated with fused deposition technology unsing a desktop 3D printer as a function of geometrical parameters of the process. *Polymers* **2018**, *10*, 313. [CrossRef] [PubMed]
10. Zhou, J.G.; Herscovici, D.; Chen, C.C. Parametric process optimization to improve the accuracy of rapid prototype stereolitography parts. *Int. J. Mach. Tools Manuf.* **2000**, *40*, 363–379. [CrossRef]
11. Naik, D.L.; Kiran, R. On anisotropy, strain rate and size effects in vat photo polymerization based specimens. *Additive Manuf.* **2018**, *23*, 181–189. [CrossRef]
12. Aslani, K.E.; Chaidas, D.; Kechagias, J.; Kyratsis, P.; Salonitis, K. Quality performance evaluation of thin walled PLA 3D Printed parts using the teguchi method and grey relational analysis. *J. Manuf Mater. Process.* **2020**, *4*, 47. [CrossRef]
13. Aznarte, E.; Ayranci, C.; Qureshi, A.J. Digital light processing (DLP): Anisotropic tensile consideration. In Proceedings of the Solid Freeform Fabrication 2017: Proceedings of the 28th Annual International Solid Freeform Fabrication Symposium: An Additive Manufacturing Conference, Austin, TX, USA, 7–9 August 2017; pp. 413–425.

 *metals*

Article

Numerical Study about the Influence of Superimposed Hydrostatic Pressure on Shear Damage Mechanism in Sheet Metals

Mohammadmehdi Shahzamanian [1,2,*], Chris Thomsen [3], Amir Partovi [1], Zhutian Xu [4] and Peidong Wu [1]

[1] Department of Mechanical Engineering, McMaster University, Hamilton, ON L8S 4L7, Canada; partovia@mcmaster.ca (A.P.); peidong@mcmaster.ca (P.W.)
[2] Department of Civil and Environmental Engineering, University of Alberta, Edmonton, AB T6G 2W2, Canada
[3] Department of Materials Engineering, NORAM Engineering and Constructors Ltd., Vancouver, BC V6C 1S4, Canada; cthomsen@noram-eng.com
[4] School of Mechanical Engineering, Shanghai Jiao Tong University, Shanghai 200240, China; zhutianxu@sjtu.edu.cn
* Correspondence: mmshahzamanian@gmail.com or mshahzam@ualberta.ca

Citation: Shahzamanian, M.; Thomsen, C.; Partovi, A.; Xu, Z.; Wu, P. Numerical Study about the Influence of Superimposed Hydrostatic Pressure on Shear Damage Mechanism in Sheet Metals. *Metals* **2021**, *11*, 1193. https://doi.org/10.3390/met11081193

Academic Editors: Antonio Mateo, Giovanni Meneghetti and Filippo Berto

Received: 6 June 2021
Accepted: 23 July 2021
Published: 27 July 2021

Publisher's Note: MDPI stays neutral with regard to jurisdictional claims in published maps and institutional affiliations.

Abstract: It is generally accepted that the superimposed hydrostatic pressure increases fracture strain in sheet metal and mode of fracture changes with applying pressure. Void growth is delayed or completely eliminated under pressure and the shear damage mechanism becomes the dominant mode of fracture. In this study, the effect of superimposed hydrostatic pressure on the ductility of sheet metal under tension is investigated using the finite element (FE) method employing the modified Gurson–Tvergaard–Needleman (GTN) model. The shear damage mechanism is considered as an increment in the total void volume fraction and the model is implemented using the VUMAT subroutine in the ABAQUS/Explicit. It is shown that ductility and fracture strain increase significantly by imposing hydrostatic pressure as it suppresses the damage mechanisms of microvoid growth and shear damage. When hydrostatic pressure is applied, it is observed that although the shear damage mechanism is delayed, the shear damage mechanism is dominant over the growth of microvoids. These numerical findings are consistent with those experimental results published in the previous studies about the effect of superimposed hydrostatic pressure on fracture strain. The numerical results clearly show that the dominant mode of failure changes from microvoid growth to shear damage under pressure. Numerical studies in the literature explain the effect of pressure on fracture strain using the conventional GTN model available in the ABAQUS material behavior library when the mode of fracture does not change. However, in this study, the shear modified GTN model is used to understand the effect of pressure on the shear damage mechanism as one of the individual void volume fraction increments and change in mode of fracture is explained numerically.

Keywords: superimposed hydrostatic pressure; shear damage growth; fracture strain; finite element analysis (FEA)

---

1. Introduction

There are several methods of increasing the ductility of metals, such as superimposing hydrostatic pressure [1,2]. In mechanical testing under superimposed hydrostatic pressure, tensile testing of the specimen is carried out in a pressure vessel that applies the desired level of pressure in the load assembly [3]. The effect of superimposed hydrostatic pressure has been studied numerically using the conventional Gurson–Tvergaard–Needleman (GTN) model under tension and bending in previous studies [1–4]. However, in this study, the modified GTN model considering the shear damage growth as an increment in the void volume fraction is used to investigate the effect of superimposed hydrostatic pressure on the shear damage mechanism.

The different failure modes shown in Figure 1 can be described in terms of the ratio of shear and normal stresses. Generally, shear failure is dominant for low or zero stress triaxiality, while the process of void nucleation, growth, and coalescence occurs when stress triaxiality is high [5]. Ashby et al. [6] investigated the influence of pressure and temperature on damage in terms of the of brittle, fully plastic, ductile, and shear fracture mechanisms. It was shown that the failure mode is a function of pressure and temperature, with an increase in pressure corresponding to an increase in ductility. It was also demonstrated that a material only fails in a fully plastic manner when all other fracture mechanisms are suppressed. This failure mode is characterized by the onset of necking that progresses to a point of zero area when a material is continuously loaded in tension past its yield point. Kao et. Al [7] used quantitative metallography to determine the effect of hydrostatic pressure on the failure mode of a steel subjected to tensile deformation. It was observed that a superimposed hydrostatic pressure suppressed the nucleation of voids and resulted in a significant increase in ductility. Unlike the void-sheet mechanism, shear decohesion is not strongly influenced by pressure; this causes the latter to be the only valid mechanism to explain the observed failure [7]. Overall, it is generally accepted that a superimposed hydrostatic pressure increases ductility by delaying or completely eliminating void nucleation and growth; this matter has been investigated in other studies [6,8–12].

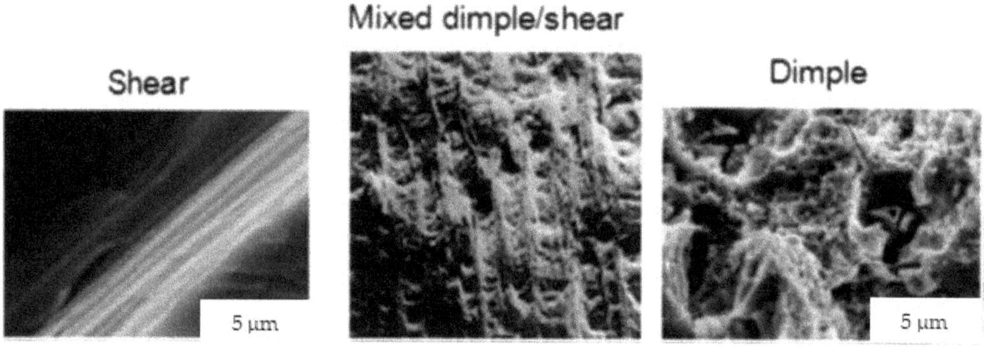

Shear

Mixed dimple/shear

Dimple

5 μm

5 μm

Figure 1. Different types of fracture mechanisms controlled by shear stress and normal stress and their combinations [5].

For many high-strength sheet materials, such as aluminum alloys that contain a significant amount of second phase particles, microvoids often develop in the vicinity of these particles during large plastic deformation. These particle-induced microvoids are known to localize plastic flow and limit the formability of sheet metals [13–16]. One of the well-known models of ductile void growth that is often utilized in analyzing large plastic deformation of ductile metallic materials is the Gurson–Tvergaard–Needleman (GTN) model, proposed by Tvergaard and Needleman [17] as an improvement on the accuracy of the original Gurson model [18]. These models treat voids as spherical cavities and capture their effects on material yield following a modification of the von Mises yield criterion [18]. More recently, the GTN model has been extended to include the effect of shear damage by Nahshon and Hutchinson [19]. Sun et al. [20] used the shear modified GTN model and simulated punch test and identified the parameters using the neural networking. The size effect on damage evolution using the shear modified GTN model under high/low stress triaxiality is performed by Li et al. [21]. Yildiz and Yilmaz [22] used the shear modified GTN model to simulate the plastic deformation for 6061 aluminum alloys. Overall, the shear modified GTN model has been used frequently to simulate various materials for different tests [23–27].

Peng et al. [4] investigated the effect of superimposed hydrostatic pressure on fracture in round bars. It was shown that, because void formation is not significant prior to necking, superimposed pressure has little or no effect on the yield strength of metals.

However, the numerical results showed that due to a suppression in void nucleation and growth by the applied pressure, the fracture strain increased, and the failure process was extended. The effect of superimposed pressure on fracture in sheet metals under tension was studied in [3], where it was again found that the application of hydrostatic pressure increased the ductility in sheet metal. Numerical results showed the transition of fracture surface from planar mode at atmospheric pressure to chisel mode under high pressure as observed experimentally.

The effect of superimposed hydrostatic pressure on the bendability of sheet metals using the GTN model in ABAQUS is investigated in [1]. This study explored how hydrostatic pressure suppresses void growth and leads to an increase in ductility in sheet metals. The pressure and stress triaxiality were shown to decrease with an increase in superimposed hydrostatic pressure. As already mentioned, the void growth decreases, and it causes the fracture strain to increase. In another study [28], the effect of cladding on the ductility of sheet metals was investigated using the GTN model. A softer material with a higher ductility than the substrate metal was applied with perfect bonding. It was demonstrated that the application of the soft ductile layer improved the bendability of the base metal. From these two studies [1,28], it is clear that combining finite element methods (FEM) with the GTN model is a useful and successful approach to perform a range of analyses and to understand various effects on the ductility of metals in three-point bending tests.

The shear damage mechanism is a dominant mode of fracture under pressure as void growth is delayed or completely eliminated. To the best of the authors' knowledge, the effect of superimposed hydrostatic pressure on the shear damage mechanism has not been reported elsewhere. The aim of this paper was to perform a numerical study of the effect of a superimposed hydrostatic pressure on shear fracture in sheet metal under tension. The effect of superimposed pressure is explained in detail and in a step-wise manner, and it is shown what happens when the shear damage mechanism becomes a dominant mode of fracture with increasing pressure. All the simulations presented in this study were performed using ABAQUS/Explicit [29] based on the modified GTN model implemented in a VUMAT subroutine. The effect of hydrostatic pressure on the change in failure mode is explained in detail. The numerical results were found to be in good agreement with experimental observations considering the mixed dimple/shear mode of fractures in a sheet metal. The void growth and shear void growth volume fractions are considered individually in the shear modified GTN model. Therefore, the effect of pressure on void growth and shear void growth volume fractions are studied and compared with each other.

2. Constitutive Model

The Gurson–Tvergaard–Needleman (GTN) model [17,30,31] is used in this study, which is on the basis of damage growth in metals due to void nucleation, growth, and coalescence. The void growth is a function of the plastic strain rate $\mathbf{D}^P$:

$$\left(\dot{f}\right)_{growth} = (1-f)\mathbf{I} : \mathbf{D}^P \tag{1}$$

and the void nucleation is assumed to be strain controlled as follows:

$$\left(\dot{f}\right)_{nucleation} = \overline{A}\dot{\overline{\varepsilon}}^P \tag{2}$$

where $\dot{\overline{\varepsilon}}^P$ is the effective plastic strain rate, and the parameter $\overline{A}$ is chosen so that nucleation follows a normal distribution as suggested by Chu and Needleman [32]:

$$\overline{A} = \frac{f_N}{S_N\sqrt{2\pi}}exp\left[-\frac{1}{2}\left(\frac{\overline{\varepsilon}^P - \varepsilon_N}{S_N}\right)^2\right] \tag{3}$$

here, f_N is the volume fraction of void nucleating particles, ε_N is the average void nucleating strain, and S_N is the standard deviation of the void nucleating strain.

Additionally, the shear damage growth proposed by Nahshon and Hutchinson [19] is as follows:

$$df_{Shear\ damage} = k_w w(\sigma_{ij}) f \frac{S_{ij} d\varepsilon^p_{ij}}{\sigma_{ef}} \tag{4}$$

k_w is the magnitude of the damage growth rate in the pure shear test. The function $w(\sigma_{ij})$ identifies the current state of stress, which is defined as $w(\sigma_{ij}) = 1.0 - \left(\frac{27 J_3}{2\sigma_{eq}^3}\right)^2$, where J_3 is the third invariant of the deviatoric stress matrix.

The growth of existing voids and the nucleation of new voids are considered in the evolution of void volume fraction as follows:

$$\dot{f} = \left(\dot{f}\right)_{growth} + \left(\dot{f}\right)_{nucleation} + \left(\dot{f}\right)_{shear\ damage} \tag{5}$$

and the function of void volume fraction $(f^*(f))$ is defined to consider coalescence as follows:

$$f^* = \begin{cases} f & for\ f \leq f_c \\ f_c + \frac{f_u^* - f_c}{f_f - f_c}(f - f_c) & for\ f > f_c \end{cases} \tag{6}$$

where f_c is the critical void volume fraction for coalescence and f_f is the void volume fraction at failure. The parameter $f_u^* = \frac{1}{q_1}$ is defined. It should be mentioned that void growth and nucleation does not happen when the stress state of an element is compressive; it may only occur in tension.

Finally, the approximate yield function to be used in which f^* is distributed randomly is as follows:

$$\Phi(\sigma, \overline{\sigma}, f) = \frac{\sigma_e^2}{\overline{\sigma}^2} + 2f^* q_1 \cosh\left(\frac{3q_2 \sigma_H}{2\overline{\sigma}}\right) - \left[1.0 + (q_2 f^*)^2\right] = 0 \tag{7}$$

where σ is the macroscopic Cauchy stress tensor and σ_e, σ_H, and $\overline{\sigma}$ are the equivalent stress, hydrostatic stress, and matrix stress, respectively. In fact, the matrix stress and equivalent stresses are damaged and undamaged stresses in the GTN model. Additionally, q_1 and q_2 are calibrated parameters.

The uniaxial elastic–plastic undamaged stress–strain curve for the matrix material is provided by the following power-law form:

$$\overline{\varepsilon} = \begin{cases} \frac{\overline{\sigma}}{E}, & for\ \overline{\sigma} \leq \sigma_y \\ \frac{\sigma_y}{E}\left(\frac{\overline{\sigma}}{\sigma_y}\right)^n, & for\ \overline{\sigma} > \sigma_y \end{cases} \tag{8}$$

3. Problem Formulation and Method of Solution

A sheet metal with length L_o, thickness t_o, and width W_o that is under hydrostatic pressure is considered and shown schematically in Figure 2. It is assumed that the sheet is wide enough and that no deformation occurs in the width direction, such that the sheet may be considered to be under plane strain. The shear modified GTN model is not supported in the ABAQUS material behavior library and a VUMAT subroutine was implemented in this study to investigate the effect of pressure on shear damage mechanism. However, the subroutine only supports the three-dimensional elements. The superimposed hydrostatic pressure is represented by small brown arrows directed into the material from all directions. The sequence of tensile strain under superimposed hydrostatic pressure is modeled as two steps. In the first step, the pressure is gradually increased up to a desired level $p = -\alpha\sigma_y$ (α defines the value of applied pressure respect with yield stress) without applying any tensile strain. In the second step, tensile strain is applied to the sheet while maintaining the constant pressure value $p = -\alpha\sigma_y$.

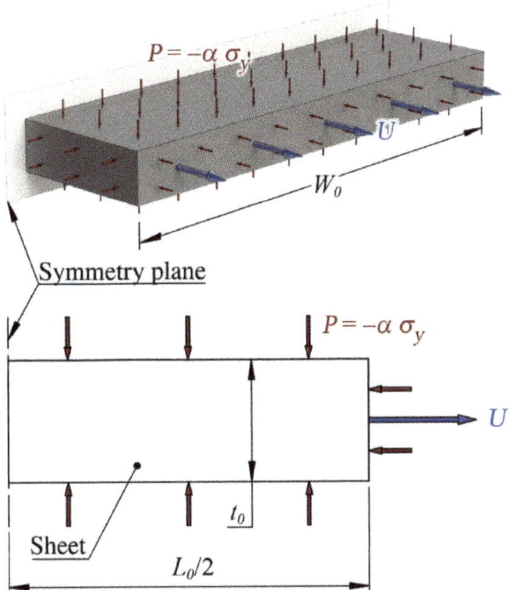

Figure 2. Schematic of a sheet metal under superimposed hydrostatic pressure.

The elastic–plastic properties of the matrix material are specified by $\sigma_y/E = 0.0033$, $\nu = 0.3$ and $n = 10$. It is assumed that the initial void volume fraction is zero and the fit parameters in the GTN model (Equation (7)) are $q_1 = 1.5$ and $q_2 = 1.0$. These values for q_1 and q_2 were found to be in good agreement in [31] for metals to analyze the bifurcation mode of porous metals. Void nucleation is assumed to be plastic strain controlled, the volume fraction of void nucleating particles $f_N = 0.04$, the mean strain for void nucleation $\varepsilon_N = 0.3$, and the corresponding standard deviation $S_N = 0.1$. The parameters related to the final failure, f_c and f_f, are assumed to be 0.15 and 0.25, respectively. These values of mechanical properties are taken from Tvergaard and Needleman [17]. It should be emphasized that the main purpose of the present study is to assess the effect of superimposed hydrostatic pressure on the ductility of sheet metals and particularly on the shear damage mechanism, and that the overall results and conclusions are not particularly dependent on the above values of the material parameters. The three-dimensional element C3D8R in ABAQUS/Explicit is used for the sheet. The mass scaling method with a sufficient low target time increment is used and it is carefully attempted to minimize the dynamic effect of the sample. Therefore, a wide sheet with a width (W_o) of 100 mm is considered when the length (L_o) and thickness (t_o) are 60 mm and 10 mm, respectively. It is to be noted that all nodes in the sheet are constrained in the width direction.

As the mesh sensitivity is expected in numerical simulations involving localized deformation and fracture, different meshes are considered in this simulation. Figure 3 shows the finite element (FE) configuration of the specimen with a typical mesh for metal sheet consisting of $60 \times 110 \times 4$ elements (60 elements in thickness direction, 110 elements in length direction and 4 elements in width direction) in which the element distribution in the refined area is biased to the middle section of the specimen where fracture is expected to occur. Due to the symmetry, only half of the sheet is investigated and symmetric boundary conditions are imposed in the middle section of specimen. Figure 4 represents the normalized force (F^*) as a function of the tensile strain ε for fully base material and the effect of mesh sensitivity on this curve is also included. Force is normalized by the multiplication of the yield stress of the material and the initial cross section of the sheet, and $\varepsilon = \ln\left(1.0 + \frac{\Delta l}{l_o}\right)$.

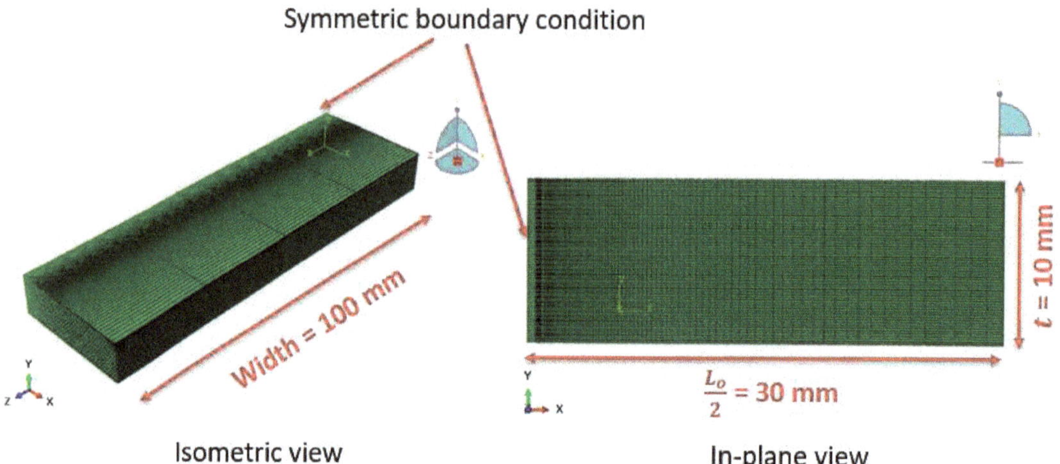

Figure 3. Finite element (FE) configuration of a specimen under tension.

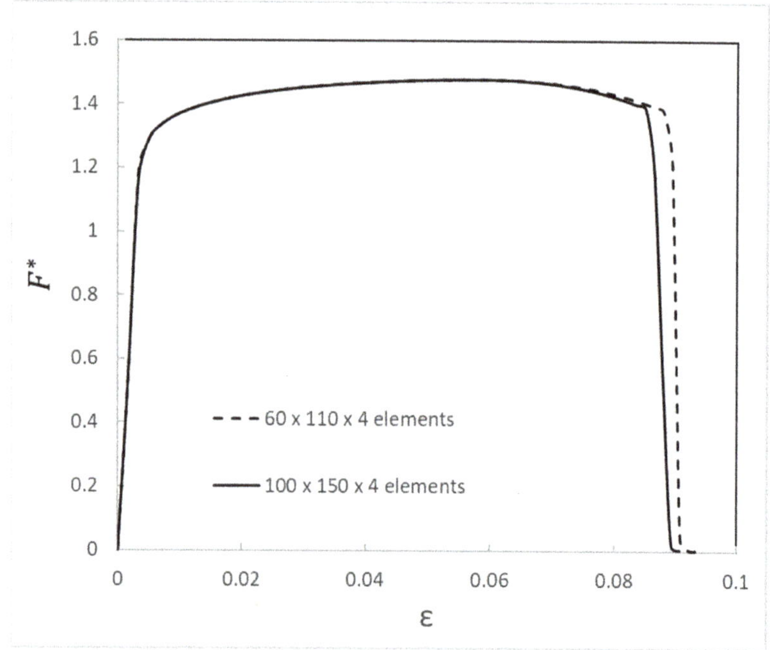

Figure 4. Effect of mesh sensitivity on force–tensile strain.

4. Results and Discussion

In this section, the $60 \times 110 \times 4$ elements mesh distribution is used to present the results. The effect of k_w on the force as a function of tensile strain under ambient pressure ($\alpha = 0.0$) is shown in Figure 5. It is clearly observed that the fracture delays with a decrease in k_w. Shear damage growth increases with an increase in k_w according to Equation (4). Therefore, the total void volume fraction increases with k_w as shown in Equation (5). It should be noted that the result for the case with $k_w = 0.0$ corresponds to using the conventional GTN model with the effect of shear damage mechanism not being considered. Additionally, the deformed shape of the fractured specimen is shown in Figure 6. Necking and localized deformation is clearly observed in the specimen. Damage is very low, close to zero, before necking and it starts to grow when localized deformation happens.

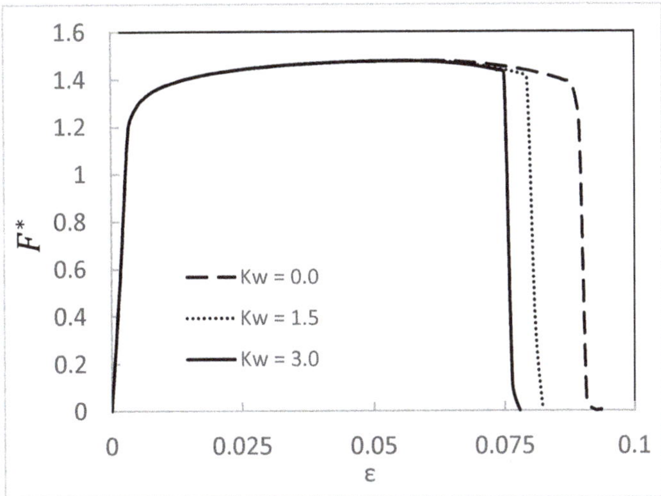

Figure 5. Effect of k_w on the normalized force–tensile strain curve.

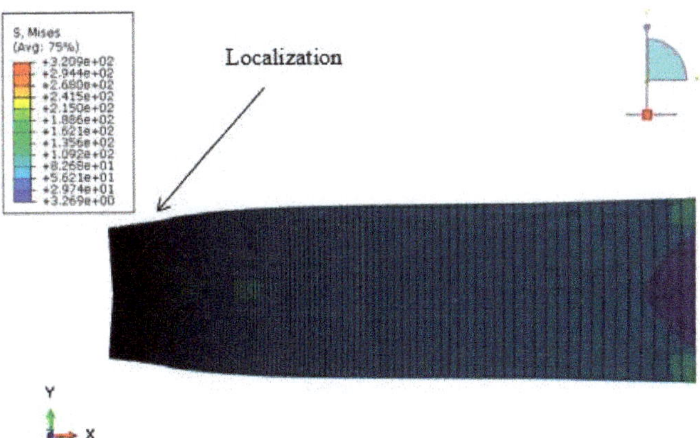

Figure 6. Deformed shape of the specimen after fracture.

As mentioned previously, the effect of pressure on ductility and bendability has been determined in [1,4] using the conventional GTN model when the effect of the shear damage mechanism is not considered. It is explained in [1,4] that the superimposed hydrostatic pressure lowers the stress triaxiality, which retards void growth and increases the fracture strain. In the present study, the influence of superimposed hydrostatic pressure $\left(p = -\alpha\sigma_y\right)$ on fracture under tension is considered while accounting for the shear damage mechanism by using the modified GTN model. Figure 7 shows the effect of α on the force–tensile strain curve. It was found that as α increases, the tensile strength is unaffected and the fracture strain of the material increases.

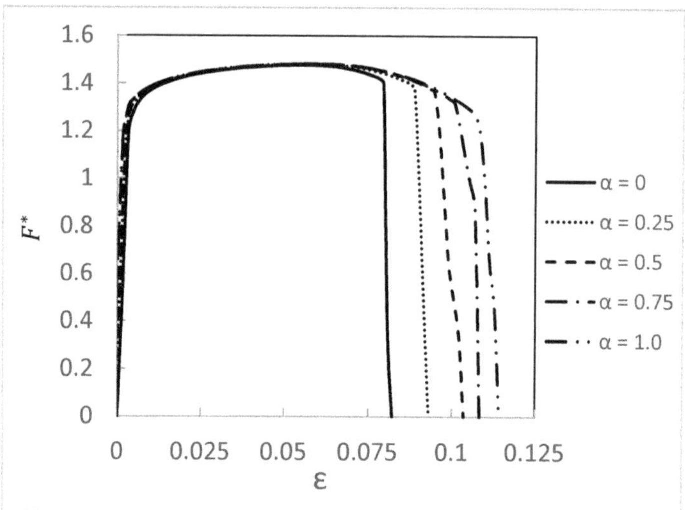

Figure 7. Effect of superimposed hydrostatic pressure on normalized force–tensile strain curves.

Figure 8 shows the volumetric strain $(\varepsilon_{11} + \varepsilon_{22} + \varepsilon_{33})$ at the center of the specimen for sheets under a range of superimposed hydrostatic pressures. It is found that the volumetric strain decreases with increasing α as shown in Figure 8. According to Equation (1), the decrease in volumetric strain renders void growth less favorable and leads to higher ductility.

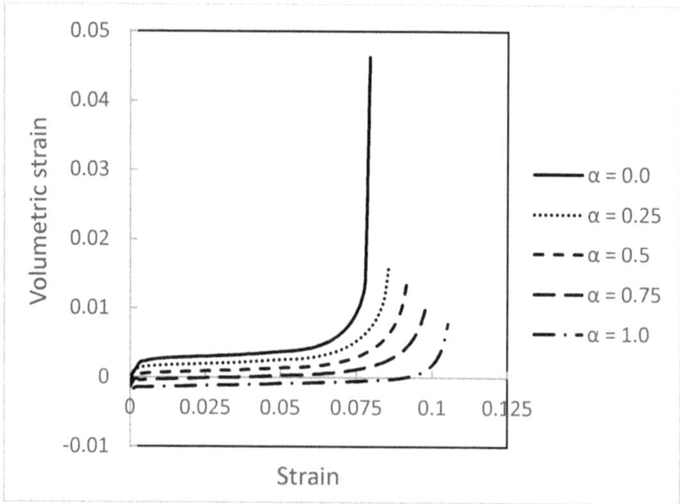

Figure 8. Effect of superimposed hydrostatic pressure on volumetric strain at the center of the specimen.

The delay of void growth and the concomitant increase in fracture strain caused by the increase in applied pressure can be explained in terms of how this pressure influences the hydrostatic pressure and stress triaxiality at the center of the specimen. Figure 9 presents the hydrostatic pressure $\sigma_H = (1/3)\left(\sigma_{xx} + \sigma_{yy} + \sigma_{zz}\right)$ and stress triaxiality $\frac{\sigma_H}{\bar{\sigma}}$ at the center of the specimen, where fracture initiates as a function of tensile strain under various superimposed hydrostatic pressures. At room pressure $p = 0$, both hydrostatic pressure and stress triaxiality develop in a way to assist void growth. However, under a superimposed hydrostatic pressure $p = -\alpha\sigma_y$, both values are initially compressive. This result implies that void growth is delayed until a sufficiently large component of tensile stress is introduced.

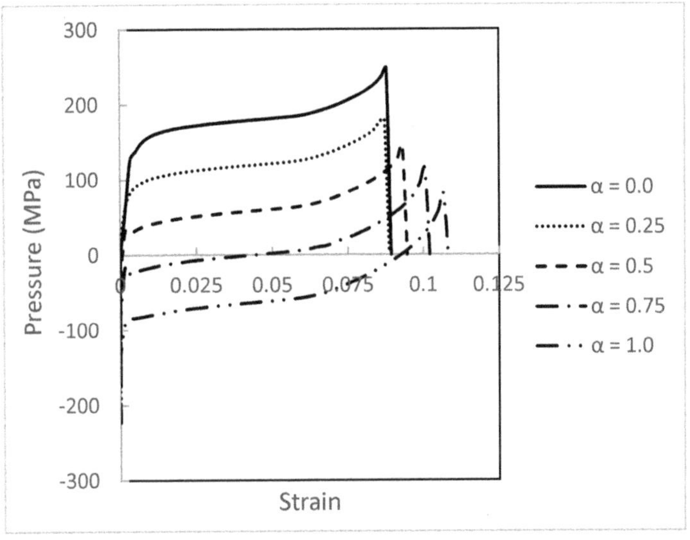

(a) Hydrostatic pressure

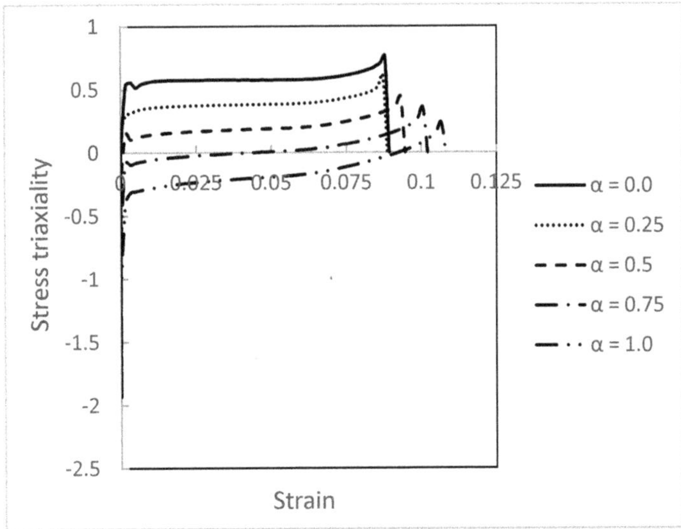

(b) Stress triaxiality

Figure 9. Effect of superimposed hydrostatic pressure on (**a**) pressure and (**b**) stress triaxiality at the center of the specimen.

The effect of superimposed double-sided pressure on the formability of a biaxially stretched age-hardenable aluminum sheet metal (AA6111-T4) was studied in [33], where the researchers numerically employed the GTN model. It was found that double-sided pressure increased formability while void nucleation remained invariable. Furthermore, only the extent of void growth was observed to change, decreasing with an increase in pressure. Figure 10a shows the effect of superimposed hydrostatic pressure on void nucleation. It is demonstrated that the final value of nucleated void volume fraction is not a function of superimposed hydrostatic pressure as the GTN model used in this study assumes that the nucleation is strain controlled (Equations (2) and (3)). A previous study [1] investigated the effect of superimposed hydrostatic pressure on bendability by using both the strain- and stress-controlled void nucleating GTN model. It was found that the final value of nucleating void volume fraction is constant when the strain void nucleating GTN model is used [1]. On the contrary, the final value of nucleating void volume fraction decreases with increasing pressure when the stress-controlled void nucleating GTN model is used. However, the effect of superimposed hydrostatic pressure on void growth is shown in Figure 10b and it is clearly seen that hydrostatic pressure delays void growth. The reduction in void growth due to an increase in the hydrostatic pressure has been reported in other studies for sheets under tension [4] and bending [28]. Figure 10c shows the effect of hydrostatic pressure on the prevalence of the shear damage mechanism and it is clearly observed that it dominates at higher values of hydrostatic pressure. As mentioned previously, the void sheet mechanism is excluded under external applied pressure, leaving shear decohesion as the dominant failure mechanism [7]. It is interesting to note that while the shear damage mechanism becomes more dominant as pressure is increased, both it and the void growth mechanism calculated using the modified GTN model become more delayed as the superimposed hydrostatic pressure is increased. Finally, Figure 10d shows the total void volume fraction under various superimposed hydrostatic pressures. It is found that the total void volume fraction delays and it will be shown that it causes the fracture strain to increase.

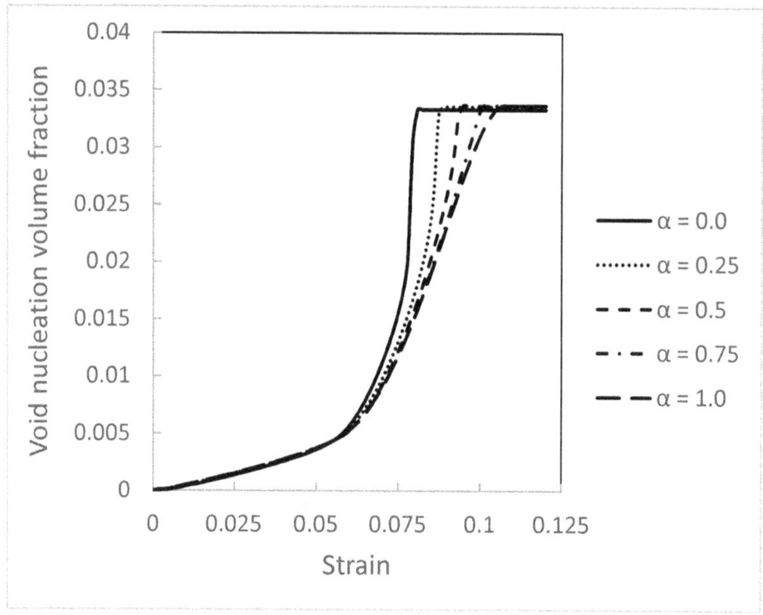

(**a**) Void nucleation volume fraction

Figure 10. *Cont.*

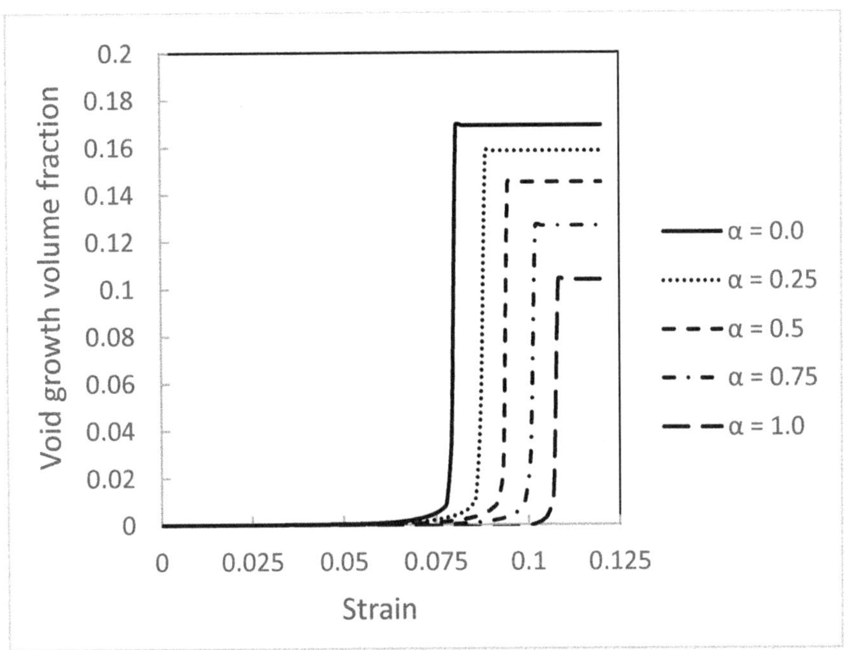

(**b**) Void growth volume fraction

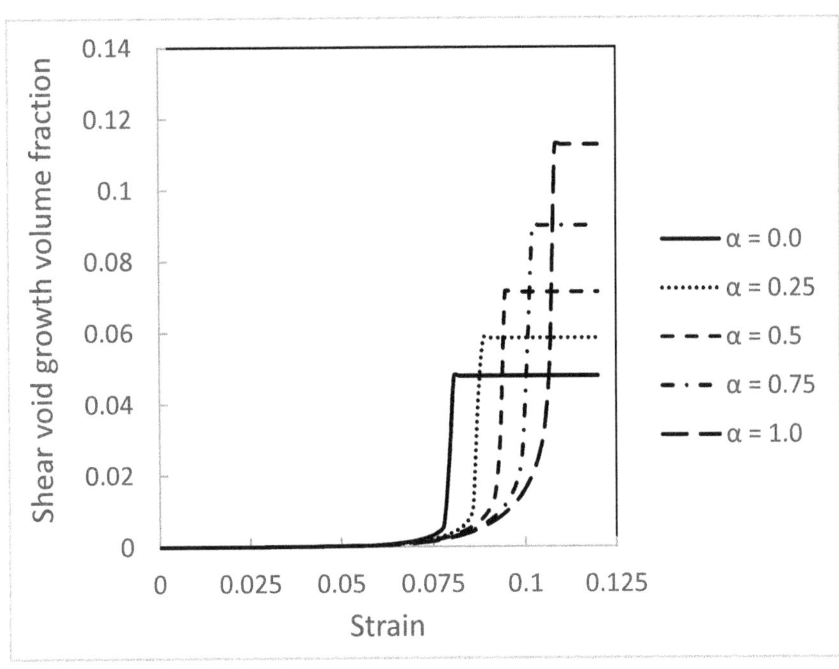

(**c**) Shear void growth volume fraction

Figure 10. *Cont.*

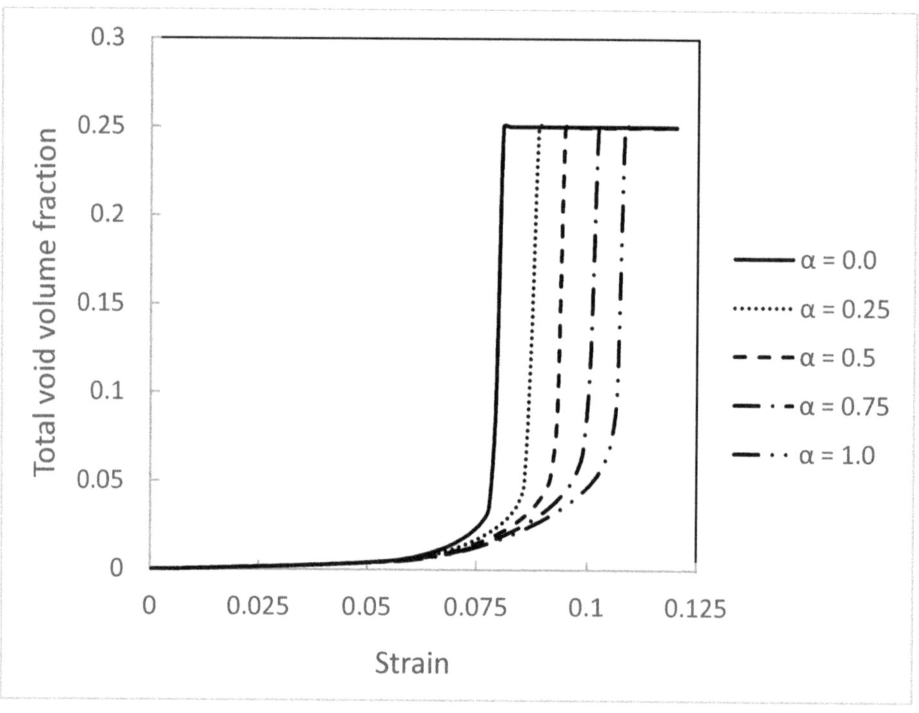

(**d**) Total void volume fraction

Figure 10. Effect of superimposed hydrostatic pressure on (**a**) void nucleation volume fraction, (**b**) void growth volume fraction, (**c**) shear void growth volume fraction and (**d**) total void volume fraction at the center of the specimen.

Figures 11 and 12 plot the influence of superimposed hydrostatic pressure on the normalized minimum cross-section area $\left(\frac{A_{min}}{A_o}\right)$ and the fracture strain $\left(\varepsilon_f\right)$ in the middle section of the specimen, respectively. Here, the fracture strain ε_f is defined as $\varepsilon_f = \ln\frac{A_o}{A_{min}}$, where A_{min} is the minimum cross-sectional area of the sheet when fracture is complete. It is to be noted that $A_{min} = t_{min}$ and $A_o = t_o$ considering the plane strain condition and in this way, $\frac{A_o}{A_{min}} = \frac{t_o}{t_{min}}$. The following equation will be obtained to calculate ε_f:

$$\varepsilon_f = \ln\frac{t_o}{t_{min}} \tag{9}$$

It is found that the minimum cross-sectional area follows an inverse relationship with the level of hydrostatic pressure. Therefore, as the pressure increases, the minimum cross-sectional area at fracture decreases and the specimen can deform more before failure, which is manifested as an increase in fracture strain.

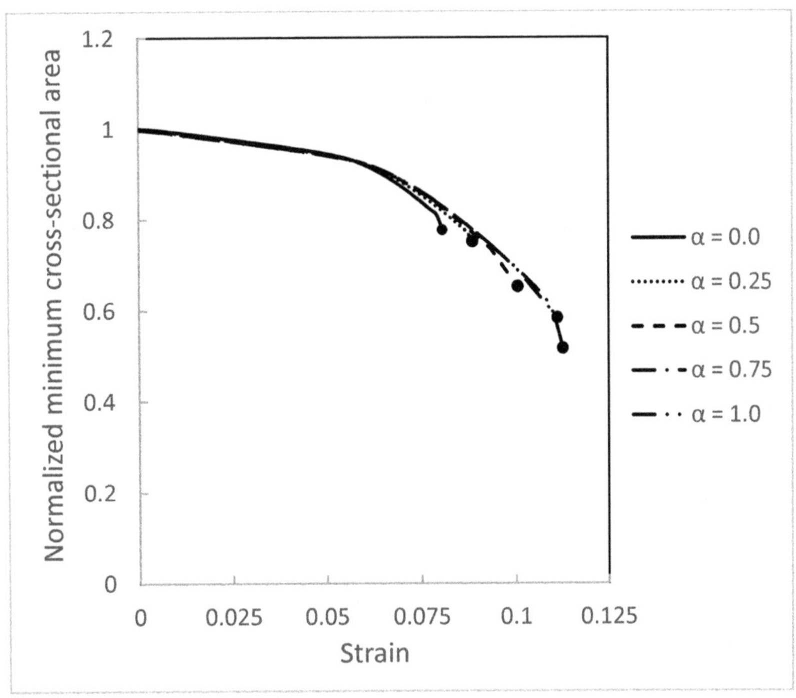

Figure 11. Effect of superimposed hydrostatic pressure on normalized minimum cross-sectional area.

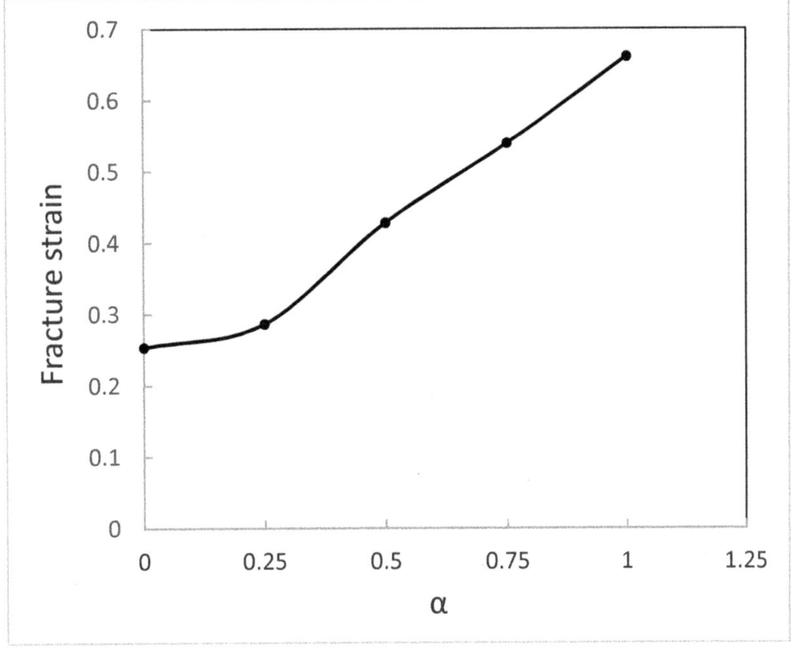

Figure 12. Effect of pressure on fracture strain.

5. Conclusions

In this study, an FEA simulation was conducted for sheet metal simultaneously subjected to a tensile test and superimposed hydrostatic pressure. The modified GTN model is used when the shear damage mechanism is considered. It is determined that the superimposed hydrostatic pressure increases the ductility significantly as hydrostatic pressure delays or eliminates growth of microvoids or microcracks as well as damage by the shear mechanism. However, it is clearly observed that the shear damage mechanism is dominant over the void growth under high pressure. The numerical results clearly show that the type of fracture changes from microvoids mechanism to shear failure under superimposed hydrostatic pressure. Finally, to sum up the conclusion remarks, the salient points are listed as follows:

- Superimposed hydrostatic pressure increases the fracture strain in metals when void growth is delayed or completely eliminated.
- Fracture mode changes under pressure and it dominates the shear damage mechanism.
- The shear modified GTN model implemented using a VUMAT subroutine explains this phenomenon when the shear damage mechanism is considered as an increment in the void volume fraction.
- Void nucleation volume fraction is constant under pressure using a strain-based void nucleating GTN model. Shear void growth volume fraction at the final fracture increases as the void growth volume fracture decreases under superimposed hydrostatic pressure.

Author Contributions: Conceptualization, M.S. and P.W.; methodology, M.S. and P.W.; software, P.W., Z.X.; validation, M.S. and P.W.; formal analysis, M.S., C.T. and P.W.; investigation, M.S., C.T. and P.W.; writing—original draft preparation, M.S., C.T., A.P., Z.X. and P.W.; writing—review and editing, M.S., C.T., A.P., Z.X. and P.W.; visualization, M.S., C.T., A.P., Z.X. and P.W.; supervision, P.W.; project administration, M.S.; funding acquisition, P.W. All authors have read and agreed to the published version of the manuscript.

Funding: This work was supported by the Natural Sciences and Engineering Research Council of Canada (NSERC, project No: RGPIN-2016-06464).

Data Availability Statement: Not applicable.

Conflicts of Interest: The authors declare no conflict of interest.

References

1. Shahzamanian, M.; Lloyd, D.; Wu, P.; Xu, Z. Study of influence of superimposed hydrostatic pressure on bendability of sheet metals. *Eur. J. Mech. A/Solids* **2021**, *85*, 104132. [CrossRef]
2. Partovi, A.; Shahzamanian, M.M.; Wu, P.D. Study of Influence of Superimposed Hydrostatic Pressure on Ductility in Ring Compression Test. *J. Mater. Eng. Perform.* **2020**, *29*, 6581–6590. [CrossRef]
3. Wu, P.D.; Chen, X.X.; Lloyd, D.J.; Embury, J.D. Effects of superimposed hydrostatic pressure on fracture in sheet metals under tension. *Int. J. Mech. Sci.* **2010**, *52*, 236–244. [CrossRef]
4. Peng, J.; Wu, P.D.; Huang, Y.; Chen, X.X.; Lloyd, D.J.; Embury, J.D.; Neale, K.W. Effects of superimposed hydrostatic pressure on fracture in round bars under tension. *Int. J. Solids Struct.* **2009**, *46*, 3741–3749. [CrossRef]
5. Kofiani, K.; Nonn, A.; Wierzbicki, T. New calibration method for high and low triaxiality and validation on SENT specimens of API X70. *Int. J. Press. Vessel. Pip.* **2013**, *111*, 187–201. [CrossRef]
6. Ashby, M.; Embury, J.; Cooksley, S.; Teirlinck, D. Fracture maps with pressure as a variable. *Scr. Met.* **1985**, *19*, 385–390. [CrossRef]
7. Kao, A.S.; Kuhn, H.A.; Richmond, O.; Spitzig, W.A. Tensile fracture and fractographic analysis of 1045 spheroidized steel under hydrostatic pressure. *J. Mater. Res.* **1990**, *5*, 83–91. [CrossRef]
8. Kao, A.; Kuhn, H.; Richmond, O.; Spitzig, W. Workability of 1045 spheroidized steel under superimposed hydrostatic pressure. *Met. Mater. Trans. A* **1989**, *20*, 1735–1741. [CrossRef]
9. French, I.E.; Weinrich, P.F. The influence of hydrostatic pressure on the tensile deformation and fracture of copper. *Met. Mater. Trans. A* **1975**, *6*, 785–790. [CrossRef]
10. French, I.; Weinrich, P.; Weaver, C. Tensile fracture of free machining brass as a function of hydrostatic pressure. *Acta Met.* **1973**, *21*, 1045–1049. [CrossRef]
11. Brownrigg, A.; Spitzig, W.; Richmond, O.; Teirlinck, D.; Embury, J. The influence of hydrostatic pressure on the flow stress and ductility of a spheroidized 1045 steel. *Acta Met.* **1983**, *31*, 1141–1150. [CrossRef]

12. Korbel, A.; Raghunathan, V.; Teirlinck, D.; Spitzig, W.; Richmond, O.; Embury, J. A structural study of the influence of pressure on shear band formation. *Acta Met.* **1984**, *32*, 511–519. [CrossRef]
13. Gologanu, M.; Leblond, J.-B.; Devaux, J. Approximate models for ductile metals containing non-spherical voids—Case of axisymmetric prolate ellipsoidal cavities. *J. Mech. Phys. Solids* **1993**, *41*, 1723–1754. [CrossRef]
14. Lievers, W.B.; Pilkey, A.; Lloyd, D. The influence of iron content on the bendability of AA6111 sheet. *Mater. Sci. Eng. A* **2003**, *361*, 312–320. [CrossRef]
15. Lievers, W.B.; Pilkey, A.; Lloyd, D. Using incremental forming to calibrate a void nucleation model for automotive aluminum sheet alloys. *Acta Mater.* **2004**, *52*, 3001–3007. [CrossRef]
16. Chen, Z.T.; Worswick, M.J.; Cinotti, N.; Pikey, A.K.; Lloyd, D. A linked FEM-damage percolation model of aluminum alloy sheet forming. *Int. J. Plast.* **2003**, *19*, 2099–2120. [CrossRef]
17. Tvergaard, V.; Needleman, A. Analysis of the cup-cone fracture in a round tensile bar. *Acta Met.* **1984**, *32*, 157–169. [CrossRef]
18. Gurson, A.L. Continuum Theory of Ductile Rupture by Void Nucleation and Growth: Part I—Yield Criteria and Flow Rules for Porous Ductile Media. *J. Eng. Mater. Technol.* **1977**, *99*, 2–15. [CrossRef]
19. Nahshon, K.; Hutchinson, J. Modification of the Gurson Model for shear failure. *Eur. J. Mech. A/Solids* **2008**, *27*, 1–17. [CrossRef]
20. Sun, Q.; Lu, Y.; Chen, J. Identification of material parameters of a shear modified GTN damage model by small punch test. *Int. J. Fract.* **2020**, *222*, 25–35. [CrossRef]
21. Li, X.; Chen, Z.; Dong, C. Size effect on the damage evolution of a modified GTN model under high/low stress triaxiality in meso-scaled plastic deformation. *Mater. Today Commun.* **2021**, *26*, 101782. [CrossRef]
22. Yildiz, R.; Yilmaz, S. Experimental Investigation of GTN model parameters of 6061 Al alloy. *Eur. J. Mech. A/Solids* **2020**, *83*, 104040. [CrossRef]
23. More, S.; Kumar, A.; Narasimhan, K. Parameter identification of GTN damage model using response surface methodology for single point incremental sheet forming of IF steel. *Adv. Mater. Process. Technol.* **2021**, 1–16. [CrossRef]
24. Li, L.; Khandelwal, K. Topology Optimization of Energy-Dissipating Plastic Structures with Shear Modified Gurson–Tvergaard–Needleman Model. *J. Struct. Eng.* **2020**, *146*, 04020229. [CrossRef]
25. Singh, A.K.; Kumar, A.; Narasimhan, K.; Singh, R. Understanding the deformation and fracture mechanisms in backward flow-forming process of Ti-6Al-4V alloy via a shear modified continuous damage model. *J. Mater. Process. Technol.* **2021**, *292*, 117060. [CrossRef]
26. He, Z.; Zhu, H.; Hu, Y. An improved shear modified GTN model for ductile fracture of aluminium alloys under different stress states and its parameters identification. *Int. J. Mech. Sci.* **2021**, *192*, 106081. [CrossRef]
27. Ying, L.; Wang, D.; Liu, W.; Wu, Y.; Hu, P. On the numerical implementation of a shear modified damage model and its ap-plication to small punch test. *Int. J. Mater. Form.* **2018**, *11*, 527–539. [CrossRef]
28. Shahzamanian, M.; Lloyd, D.; Wu, P. Enhanced bendability in sheet metal produced by cladding a ductile layer. *Mater. Today Commun.* **2020**, *23*, 100952. [CrossRef]
29. *Abaqus/CAE User's Manual*; ABAQUS Inc.: Waltham, MA, USA, 2014.
30. Tvergaard, V. Influence of voids on shear band instabilities under plane strain conditions. *Int. J. Fract.* **1981**, *17*, 389–407. [CrossRef]
31. Tvergaard, V. On localization in ductile materials containing spherical voids. *Int. J. Fract.* **1982**, *18*, 237–252.
32. Chu, C.C.; Needleman, A. Void Nucleation Effects in Biaxially Stretched Sheets. *J. Eng. Mater. Technol.* **1980**, *102*, 249–256. [CrossRef]
33. Liu, J.; Wang, Z.; Meng, Q. Numerical Investigations on the Influence of Superimposed Double-Sided Pressure on the Formability of Biaxially Stretched AA6111-T4 Sheet Metal. *J. Mater. Eng. Perform.* **2011**, *21*, 429–436. [CrossRef]

metals

Article

Reduction of Warping in Kinematic L-Profile Bending Using Local Heating

Eike Hoffmann *, **Rickmer Meya** and **A. Erman Tekkaya**

Institute of Forming Technology and Lightweight Components (IUL), TU Dortmund University, Baroper Str. 303, 44227 Dortmund, Germany; rickmer.meya@iul.tu-dortmund.de (R.M.); erman.tekkaya@iul.tu-dortmund.de (A.E.T.)

* Correspondence: eike.hoffmann@iul.tu-dortmund.de; Tel.: +49-0231-755-6926

Abstract: Kinematic bending of profiles allows to manufacture parts with high flexibility concerning the geometry. Still, the production of profiles with asymmetric cross-sections regarding the force application axis using kinematic bending processes offers challenges regarding springback and warping. These geometric deviations can be reduced by partial, cross-sectional heating during the process as it lowers the flow stress locally. In this work, the influence of partial, cross-sectional heating during a three-roll push-bending process on the warping and springback of L-profiles is investigated. Numerical and experimental methods reveal the influence of temperature on warping and springback. A newly developed analytical model predicts the warping and bending moment in the design phase and assists to understand the effect of warping reduction through partial heating during plastic bending. With increasing temperature of the heated profile area, the warping is reduced up to 76% and the springback of the bend profiles is decreased up to 44%. The warping reduction is attributed to a shift in stress free fiber due to the temperature gradient between heated and room temperature areas. The shift of stress-free fiber leads to an adapted shear center position, resulting in an approximated "quasi-symmetric" bending case.

Keywords: kinematic bending; product properties; local heating; profile bending; asymmetric profile; warping

Citation: Hoffmann, E.; Meya, R.; Tekkaya, A.E. Reduction of Warping in Kinematic L-Profile Bending Using Local Heating. *Metals* **2021**, *11*, 1146. https://doi.org/10.3390/met11071146

Academic Editor: José Valdemar Fernandes

Received: 24 June 2021
Accepted: 16 July 2021
Published: 20 July 2021

Publisher's Note: MDPI stays neutral with regard to jurisdictional claims in published maps and institutional affiliations.

1. Introduction

The warping of profiles with an asymmetric cross-section regarding the force application axis during bending is a common issue in profile bending. The shift between the force application axis and the shear center induces torsion moments in the cross-section [1] (Figure 1a). Altering the position of the force application axis to the shear center either requires specialized tools or is not possible without changing the product geometry as the position of the shear center can be located outside of the profile cross-section [2]. However, it is possible to move the force application axis to the shear center through welded plate elements (see Figure 1b). Methods to counter warping include the superposition with counteracting torsion-inducing forces [3]. One example of this method is the torque superposed spatial bending (TSS) process [4]. In this process, additional torque is superposed during the bending process which suppresses profile warping. The disadvantage is the additional torque axis necessary. Other methods to suppress warping include the use of supporting elements [5]. In a patent by Kreye [6], plastic warping is preventable through geometrical restrictions by supporting the profile at the whole shell surface in the forming area (see Figure 1c). Support elements have also been used to suppress the formation of a secondary bending axis in the bending of asymmetric profiles through a flexible bending process [7]. In the aforementioned flexible bending process, the profile is guided through a pusher. The position of the pusher is adjustable in the profile length direction which changes the position of the forming zone.

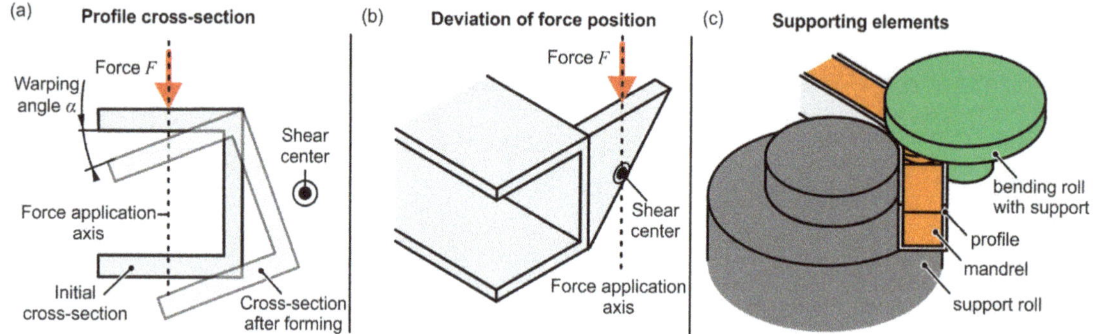

Figure 1. Countermeasures to reduce warping in profile bending profiles for the example of a U-profile. (**a**) profile cross-section with the shear center as cause for warping, (**b**) countermeasure: moving the position of force application by altering the profile geometry. (**c**) countermeasure: use of supporting elements.

Additionally, process control can be used to counteract geometric deviations in kinematic bending processes. Wang et al. [8] proposed a process control concept by asymmetric loading patterns in a four-roll bending process of Z-profiles. Through the asymmetric loading pattern, warping can be eliminated. In return, process control requires complex models and high computational efforts. The mentioned methods to decrease geometric deviations in profile bending reduce flexibility as either more tools, machine axes, or workpiece-dependent models for process control are necessary. To reduce geometric deviations in kinematic profile bending, other methods are required to maximize flexibility.

To decrease the number of geometric deviations in bending processes without workpiece-dependent measures, bending processes have been supported by heating methods [9]. Park et al. [10] managed to employ a synchronous incremental forming and incremental heating process to bend hat-shaped profiles consisting of DP590 without buckling, collapse, or necking. Behne [11] describes a bending process with local induction heating and consequent cooling for tubes. The process manages to reduce thinning for bending radii of up to 2.5 times the diameter without a mandrel. Yanagimoto et al. [12] managed bending without springback for the v-bending of high-strength steel sheets. The reduction of springback can be attributed to the lowered bending moments due to thermal softening.

To use heating as a warping reduction measure, a flow stress gradient in the profile cross-section is to be reached through heating of one area of the cross-section, while the temperature of other areas remains approximately at room temperature. Local, cross-sectional heating is a known concept to reduce geometrical deviations in profile products. The so called "heat-straightening" is used in bridge [13] or ship repair [14] to suppress deformations due to thermal or mechanical loads. In this process, heat is introduced into a deformed part. Thermal expansion of a heated part is restricted to build up compressive stresses. Upon cooling, the amount of plastic compression remains. Through this process, deformed parts can be restored to their original shape. This method is not integrated into manufacturing processes and the knowledge is mostly empirical. The development of laser bending evolved from this method. Laser bending uses heat generated by a laser to induce thermal stresses in a workpiece, which leads to the shaping of the part. Kraus [15] used a wedge-shaped heating strategy to realize the bending of rectangular tubes. The products produced in this process are not subject to springback. In the laser bending of tube parts, it is shown that the process reduces tube ovalization compared to mechanical bending without a mandrel [16]. While the laser bending process offers advantages compared to conventional bending, a complex heating strategy is necessary to produce bend profiles without distortion.

The state of the art shows various methods to decrease warping in the bending of profiles and methods to decrease geometrical deviations in bending by local heating. On the one hand, these methods require special tools or computational efforts to reduce warping and springback. On the other hand, the displayed methods are not tested on profiles with asymmetry to the force application axis. In this work, a novel heating strategy by partial, cross-sectional heating and cooling to support a three-roll push-bending process is analyzed. The heating strategy aims to reduce profile warping and springback of the products. A gradient stiffness over a profile cross-section will influence the position of the shear center [17]. Additionally, through heating of the profile part, the flow stress and Young's modulus, and therefore the stiffness, will decrease. The decreased flow stress will also lead to lowered stresses in the profile in this area. Both of these effects will consequently influence the shear center position. As the flow stress gradient will also cause a shift in stress free fiber position, it is suspected that the shear center position and the stress-free fiber position are linked. Through the decreased flow stress, a springback reduction in the bending process is expected.

Three main hypotheses emerge from the observations in the literature:

1. Partial heating while continuously bending an asymmetric profile leads to reduced springback.
2. Warping is reduced through partial, cross-sectional heating in kinematic bending processes of asymmetric profiles.
3. Warping reduction by partial, cross-sectional heating in the bending of asymmetric profiles is correlated with the stress-free fiber position.

To prove the defined hypotheses, L-profiles of S500MC are tested in a three-roll push bending process with partial-cross sectional heating and consequent water-jet cooling. The resulting unloaded profiles are evaluated for springback and warping and are compared to FEM results. With the geometrically validated FEM results in the unloaded state, the model in the continuous push-bending phase is used to describe the bending moment, warping, and position of stress-free fiber during the process. These results are compared to the developed analytical model for the continuous push-bending phase, which is able to predict the profile warping and bending moment.

In Table 1 the Nomenclature for this manuscript is presented.

Table 1. Nomenclature.

b	Profile width	[mm]
C_i	Material constant	/
d_{rl}	Lower roll diameter	[mm]
d_{ru}	Upper roll diameter	[mm]
E	Young's modulus	[GPa]
E_p	Mean slope of the flow curve	[MPa]
F	Force	[N]
I_z	Second moment of area	[mm^4]
l_{bend}	Distance to bending roll	[mm]
l_{cool}	Distance to cooling zone	[mm]
l_{heat}	Distance to heating area	[mm]
l_p	Distance to onset of plasticity	[mm]
M_b	Bending moment	[Nm]
M_T	Torsion moment	[Nm]
Q	Cross sectional force	[N]
R_m	Tensile strength	[MPa]
$R_{p0,2}$	Yield strength	[MPa]
S	First area of moment	[mm^3]
r_m	Bending radius	[mm]
r_{mC}	Bending radius at profile center fiber	[mm]
r_i	Bending radius at profile inner fiber	[mm]
r_{iR}	Unloaded radius at profile inner fiber	[mm]

Table 1. *Cont.*

t	Profile thickness	[mm]
T	Temperature	[°C]
T_{heat}	Heating temperature	[°C]
T_{RT}	Room temperature	[°C]
k_f	Flow stress	[MPa]
v_f	Feed velocity	[mm/s]
x, y, z	Coordinate axis	[mm]
x_0, y_0	Distance from cross-section center to center of gravity	[mm]
y_f	Strain free fiber position	[mm]
y_m	Stress free fiber position	[mm]
y_{pl}	Fiber position of plasticity onset	[mm]
α	Warping angle	[°]
a_T	Coefficient of thermal expansion	[1/K]
ε_{el}	Elastic strain	/
ε_{pl}	Plastic strain	/
$\bar{\varepsilon}_{pl}$	Plastic equivalent strain	/
$\dot{\bar{\varepsilon}}_{pl}$	Plastic equivalent strain rate	[1/s]
$\dot{\bar{\varepsilon}}_{pl,0}$	Initial plastic equivalent strain rate	[1/s]
θ	Bending angle	[°]
ν	Poisson ration	/
σ	Normal stress	[MPa]
σ_B	Bending stress	[MPa]
τ	Shear stress	[MPa]

2. Materials and Methods

2.1. Process Setup and Process Parameters

For the analysis of the warping and springback, bending experiments are conducted. The experimental set-up is executed as a kinematic three roll push-bending process (Figure 2). The process is divided in three phases: pre-bending, kinematic push bending, and unloading. In pre-bending (Figure 2a), the profile is clamped between guide roll, counter roll, and bending roll. Between guide and counter roll the profile is only guided so no force is implied in profile thickness direction. To initiate the bending force between bending and counter roll, the bending roll is rotated until the desired bending radius is reached. During the kinematic push-bending phase (Figure 2b) the profile is pushed forward with the feed velocity v_f to reach the desired bending angle. The profile is partially heated by induction in one profile area and consequently cooled by a water jet to keep the heated zone small with a length of the heated zone of 95 mm during this phase. Heating and cooling are deactivated for the unloading phase (Figure 2c). The bending roll is retracted, and the profile will spring back.

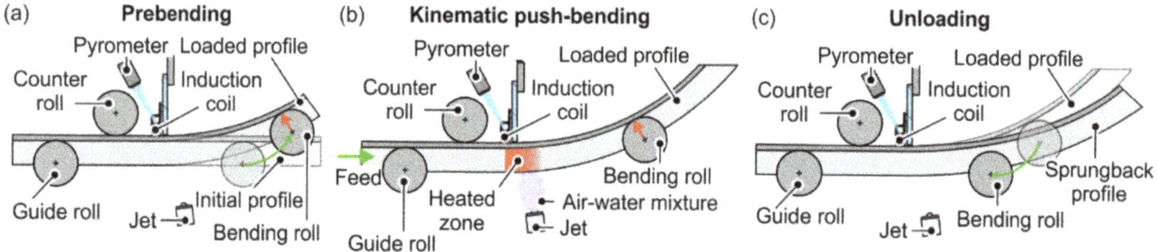

Figure 2. Process concept for (**a**) prebending phase, (**b**) kinematic push-bending phase, and (**c**) unloading phase.

The experiments are performed on a rotatory drawbending machine DB 2060-CNC-SE-F, built by transfluid, Schmallenberg, Germany (Figure 3). The temperature is measured by a pyrometer M318 produced by Sensortherm, Steinbach, Germany. The temperature is measured at the center of the profile width after the profile passes the induction coil. Based on the measured temperature, the power output of the induction generator is controlled. Heating of the profiles is achieved through an induction generator TruHeat 7040 manufactured by Trumpf Hüttinger, Freiburg, Germany with a power of 40 kW. The heating and cooling zones are divided by a sheet of Dotherm D800-M, produced by Moeschter group, Dortmund, Germany, which possesses high temperature stability and low conductivity, to prevent water from reaching the cooling area. The geometry of the unloaded profiles is digitalized through the 3D scanning system ATOS, manufactured by GOM, Braunschweig, Germany, to quantify springback and warping.

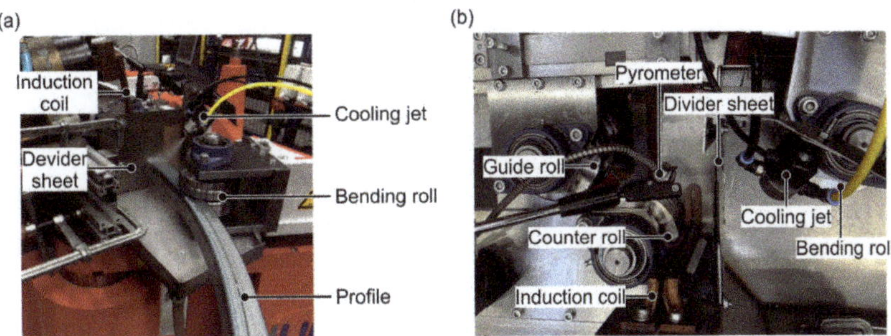

Figure 3. Experimental setup, (**a**) frontal view, (**b**) top view between the rolls.

L-profiles with a width b of 40 mm and a thickness t of 2.5 mm are analyzed. The loaded bending radius is set to 600 mm. The profile feed velocity is varied between 4 mm/s and 8 mm/s to show the influence of feed velocity on springback. The temperatures for the partial heating of a single profile area are room temperature and varying temperatures between 200 °C to 600 °C in 100 °C steps. Higher temperatures are not investigated as material recrystallization mechanisms are not considered in this work. The term partial heating always refers to the heating of the area of the L-profile parallel to the force as depicted in Figure 2b).

2.2. Principal Forming Zone

Depending on the partial heating temperature of the profile, a deviation of the forming zone position from the counter roll position is possible. The forming zone is at the position the bending moment in the profile first surpasses a threshold bending moment. At room temperature ($T = T_{\mathrm{RT}}$), the flow stress for forming is higher than in the heated case (T_{heat}) (Figure 4a). The flow stress results in a threshold bending moment that is necessary to achieve a bending radius of r_{m} (Figure 4b). One area of the profile is at room temperature (T_{RT}) while the other is at heated temperature (T_{heat}) for the partially heated profile. As the bending moment is an integral behavior demonstrated over the cross-section geometry, the threshold moment is decreased for a partially heated profile compared to a profile that is at room temperature. This results in the formation taking place at the counter roll position for the room temperature case. Partial heating for this process does not take place directly under the counter roll, but at a position $x = l_{\mathrm{heat}}$ due to technical limitations. As the threshold bending moment is decreased between $x = l_{\mathrm{heat}}$ and $x = l_{\mathrm{cool}}$, plasticity would not necessarily start at the counter roll. If the threshold moment for the heated area is reached, plasticity starts in the heated zone ($x = l_{\mathrm{heat}}$). The position at which plasticity starts $x = l_{\mathrm{p}}$ is consequently either the position of the counter roll ($l_{\mathrm{p}} = 0$) or the position of the heated area ($l_{\mathrm{p}} = l_{\mathrm{heat}}$). The position of the forming zone is dependent on the amount of flow stress reduction and the position of the heated area.

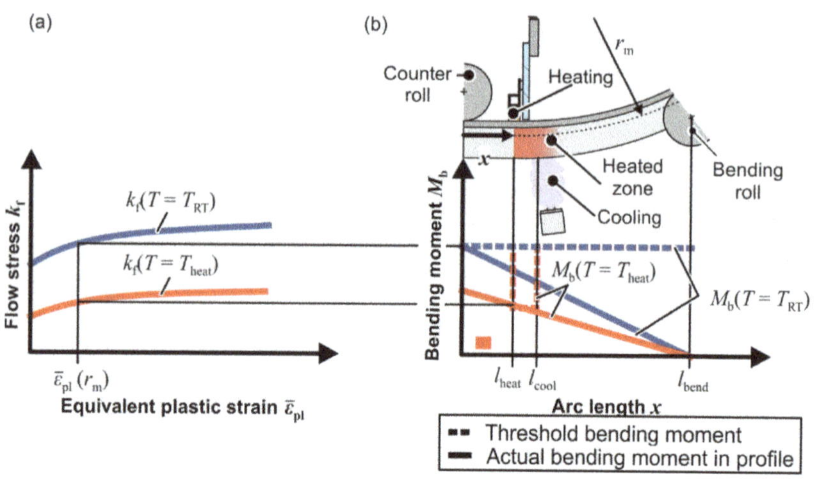

Figure 4. (**a**) Resulting flow stress k_f to bend a profile to the desired radius r_m, (**b**) necessary bending moment to reach the desired radius and actual profile bending moment for room temperature (Mb (T_{RT})) and partial heated case (Mb (T_h)).

2.3. Numerical Model and Material Parameters

The material in use for the profiles is S500MC. Delivery conditions are according to EN 10140-2 (Table 2).

Table 2. Mechanical properties and chemical composition of steel S500MC in delivery condition according to EN 10140-2.

Yield Strength $R_{p0,2}$ in MPa	Tensile Strength R_m in MPa	Chemical Composition in wt%								
		C	Si	Mn	P	S	Al	Nb	Ti	V
585	642	0.045	0.02	0.812	0.013	0.007	0.032	0.013	0.001	0.137

The temperature-dependent Young's modulus and temperature- and strain rate-dependent flow curves were characterized through isothermal tensile tests on a Z250 tensile testing machine built by ZwickRoell, Ulm, Germany with induction heating. For the tensile tests, specimens with a parallel length of 30 mm and width of 10 mm are used to achieve a homogenous heating. The specimens are produced through laser cutting of the profile material in delivery condition. The strains are measured using a Maytec PMA-12/1N7-1 Extensometer produced by Maytec, Singen, Germany. Logarithmic strain rates 0.0003, 0.003, 0.03, 0.3, and 0.1 1/s as well as temperatures from 25 °C and between 200 and 600 °C in 100 °C steps were investigated (Figure 5). The specimens are evaluated according to ISO 6892. The software Abaqus 2018 with explicit global time incrementation is used for the numerical FE simulation. The profile is discretized with tri-linear hexahedral elements with full integration (C3D8T). The simulations employ full thermo-mechanical coupling. The deformable profile is meshed with 16 elements over the profile width (2.5 mm thick elements). Isotropic, linear, temperature-dependent elasticity is assumed for the workpiece material. The plastic behavior is modelled as isotropic according to von Mises with temperature and strain-rate dependent hardening. The rolls are modelled as rigid bodies with temperature degrees of freedom. The flow curves for each temperature set are extrapolated using a Tanimura–Voce model (see Appendix A for the model parameters)

$$k_f = C_1 + (C_2 - C_1) \exp\left(-C_3 \cdot \bar{\varepsilon}_{pl}\right) + \left(C_4 - C_5 \cdot \bar{\varepsilon}_{pl}{}^{C_6}\right) \log\left(\frac{\dot{\bar{\varepsilon}}_{pl}}{\dot{\bar{\varepsilon}}_{pl,0}}\right) + C_7 \dot{\bar{\varepsilon}}_{pl}{}^{C_8} \qquad (1)$$

with the flow stress k_f, the material constants C_1 to C_8, the equivalent plastic strain $\bar{\varepsilon}_{pl}$, the equivalent plastic strain rate $\dot{\bar{\varepsilon}}_{pl}$, and the initial equivalent plastic strain rate $\dot{\bar{\varepsilon}}_{pl,0}$. The temperature dependency of Young's modulus is approximated by

$$E(T) \approx -0.1579\frac{\text{MPa}}{\text{K}^2} \cdot \Delta T^2 - 9.5596\frac{\text{MPa}}{\text{K}} \cdot \Delta T + 180,098 \text{ MPa}, \tag{2}$$

where ΔT is the difference between room temperature and heating temperature. The maximum error for this approximation is 6%. The Poisson's ratio is set as 0.3.

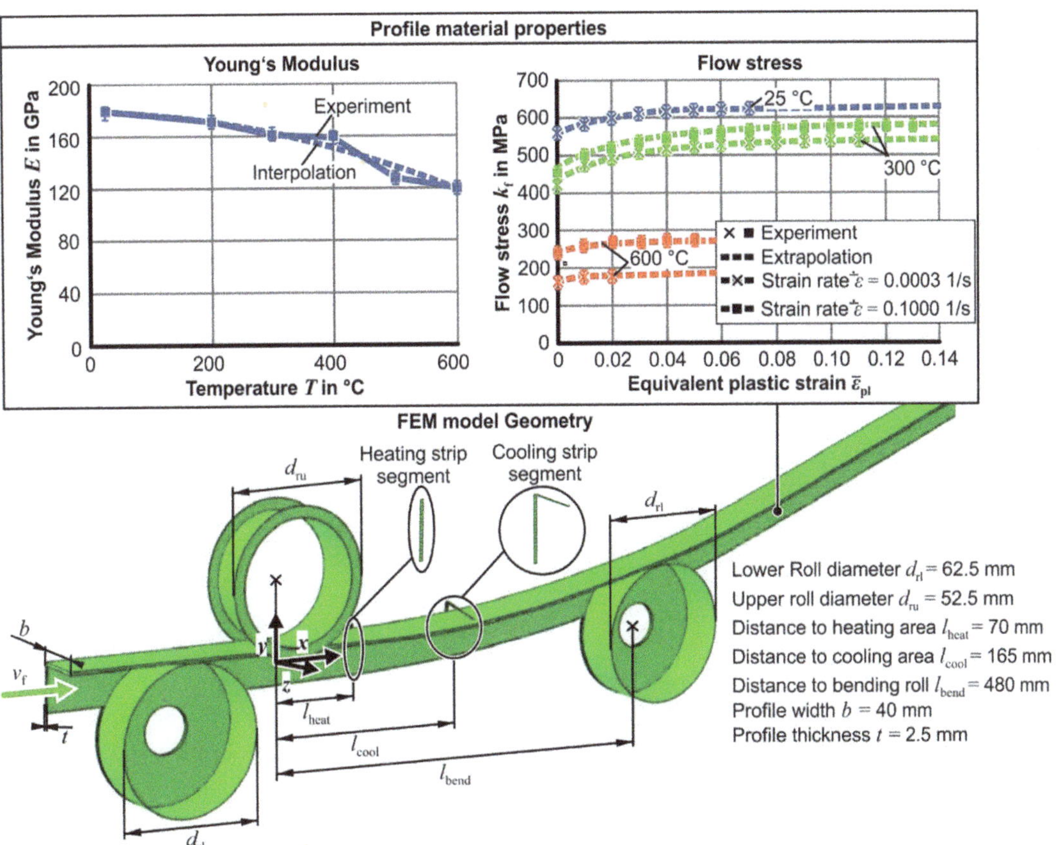

Figure 5. Simulation set-up with geometrical- and material parameters.

Continuous heating and cooling are realized through isothermal rigid strips with high conductivity. This heating method assumes that the temperature is evenly distributed after the material reaches the heating or the cooling zone. The strip for the heated area is at maximum temperature while the strip for the cooling area is at room temperature. The simulation is divided into three steps comparable to the bending tests. These steps are pre-bending, temperature-assisted continuous push bending, and unloading.

3. Analytical Modelling of the Kinematic Push-Bending Phase

In this segment, the developed analytical procedure is explained. First, the process is abstracted for the simplification of the calculations. Then, the assumptions for the model are displayed. After that, the resulting equations for strains and strain rates are defined.

Using these definitions, the methods to evaluate warping angle and bending moment are shown.

3.1. Abstraction of the Process Geometry Regarding the Profile Load

Plasticity in the analytical model is assumed to start at a position $x = l_p$ to reach a set bending radius at the center fiber r_{mC} for the profile (see Figure 6a). The counter roll is simplified as fixed support, at which the origin of the coordinate system is located in the profile cross-sectional center of gravity. Resulting from the bending, an angle θ is generated between the forming force F and the position of plasticity onset l_p. The forming force F induces a bending moment M_b. The partial heating on the profile is assumed to shift the position of the stress-free fiber y_m (T). Additionally, through the forming force F, a section force Q on the cross-section (Figure 6b) is implied, which results in shear stresses τ_{yz} due to the difference between the force application axis position and shear center position. The shear stresses result in a torsion moment M_T leading to a warping deformation with the warping angle α. The alteration of the stress-free fiber position y_m (T) is assumed to influence the position of the shear center.

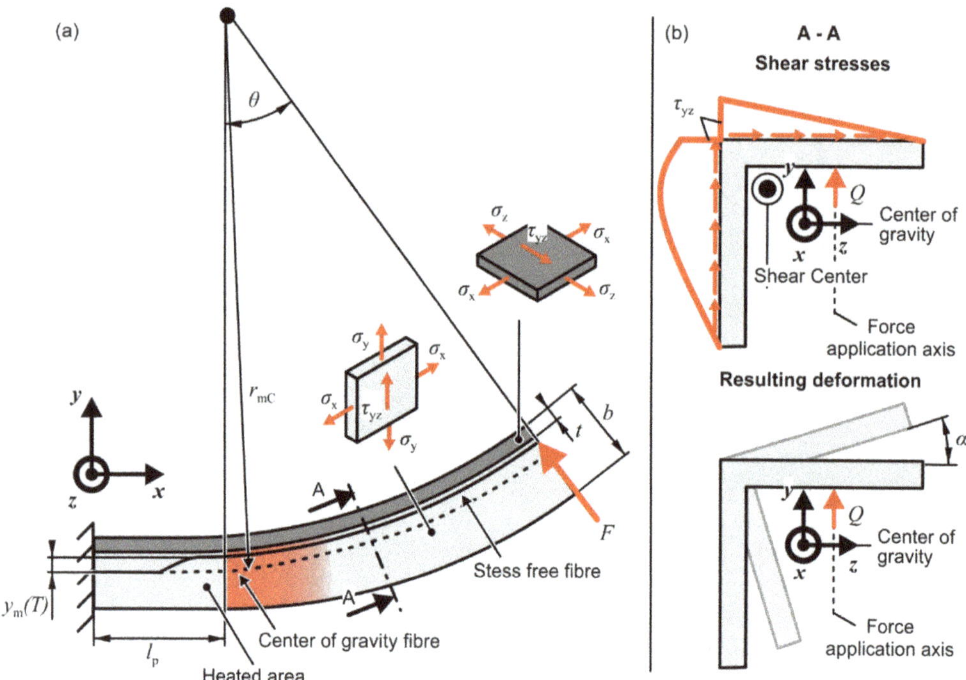

Figure 6. Mechanical model for analytical calculation. (**a**) length direction, (**b**) corresponding cross-sectional cut A-A.

3.2. Assumptions

For the analytical investigation, it is assumed that a modified version of the elementary bending theory is applicable. The assumptions used from elementary bending theory are [18]

1. Bending radius r_{mC} is constant during the whole bending operation
2. Planar surfaces stay planar and vertical to the sheet surface
3. The material is homogenous and isotropic
4. The sheet thickness during the bending process is constant Deviating from elementary bending theory, the following assumptions are made:

5. Normal stress implied by bending force is negligible compared to the bending and shear stresses
6. The progression of stress and strain curves for the tensile and the compressive area is symmetrical to the center of gravity fiber if the whole profile is at room temperature
7. The stress state for each of the cross-sectional areas is planar (elementary bending theory for wide sheets) with an addition of shear stresses vertical to the profile thickness due to the difference in position of force application axis and shear center.
8. The pre-bending phase has no influence on process behavior during the continuous push-bending phase.
9. The influence of shear stresses on reaching the threshold of onset of plasticity is negligible.
10. Through constant feed velocity v_f the process is stationary at position x
11. Due to the small distance between the heating and the cooling zone no exchange of heat between the heated profile area and the area at room temperature is considered and heat exchanged with the surroundings is neglected (see Appendix B).
12. In the heating area, temperature increases linearly until reaching a maximum temperature at the center of the heating area.

3.3. Calculation of Strains and Strain Rates

In the continuous push bending process, the material behaves elastically at $x < l_p$ and the material can behave partially elastic or fully plastic over the cross-section. The temperature dependent elastic bending strain $\varepsilon_{el,x}$ can be described as

$$\varepsilon_{el,x} = \frac{-y}{r_{mC}} + a_T \Delta T, \tag{3}$$

with the loaded bending radius to the center of gravity fiber r_{mC}, the thermal expansion coefficient a_T and the difference between room temperature and heating temperature ΔT as the compressive zone is in the positive y-segment. Elastic shear deformation is neglected ($\varepsilon_{el,yz} = 0$) Using Equation (3), the strain free fiber position y_f can be calculated through $\varepsilon_{el,x}(y_f) = 0$. For the plastic area of the profile cross-section, the plastic strains are expressed as

$$\varepsilon_{pl,x} = \ln\left(1 + \frac{-y}{r_{mC}}\right), \tag{4}$$

$$\varepsilon_{pl,yz} = \tan(\alpha). \tag{5}$$

Elastic and thermal strain components are neglected if the material behaves plastic as they are considered small compared to the plastic mechanical strain (see Appendix C).

The bending strain rate in the continuous push-bending phase can be expressed as

$$\dot{\varepsilon}_{pl,x} = \frac{v_f}{x - l_p} \ln\left(1 + \frac{-y}{r_{mC}}\right), \tag{6}$$

with the profile feed velocity v_f and the x position of plasticity onset l_p. Assuming that strain rates and strains at the position x remain constant over time, during the bending, due to constant feed velocity, the Levy–Mises flow rule can be used to calculate the shear strain rate in the plastic segment as

$$\dot{\varepsilon}_{pl,yz} = \varepsilon_{pl,yz} \frac{\dot{\varepsilon}_{pl,x}}{\varepsilon_{pl,x}}. \tag{7}$$

3.4. Calculation of Profile Warping

Using the mean slope E_p (T) (Appendix D) of the flow curve defined in Equation (1) and using temperature dependent Young's modulus definition (Equation (2)), a relation between shear stress and shear strain rate in the plastic zone can be described through constant temperature as [19]

$$\tau_{yz} = \frac{1}{3} \frac{E(T) \cdot E_{\mathrm{p}}(T)}{E(T) + E_{\mathrm{p}}(T)} \tan(\alpha).$$ (8)

To describe the correlation between shear strains and forming force, the equation

$$\tau_{yz} = \frac{Q \cdot S}{I_z \cdot t}$$ (9)

is used with the cross-sectional force Q, the first moment of area S, the second moment of area I_z and the profile thickness t. Using the relation between torsion moment and shear stress

$$M_T = \int \tau_{yz} \cdot t \, dA,$$ (10)

(see Appendix E for the formulated integral) and using Equations (8) and (9) the warping angle α can be calculated numerically.

3.5. Calculation of the Bending Moment

The equation for the bending moment of the process can be expressed as

$$M_b = \int \sigma_{\mathrm{B}} \cdot y \, dA,$$ (11)

where the bending stress in the plastic area $\sigma_{\mathrm{B,pl}}$ can be resolved through the Mises equation and the flow rule

$$\sigma_{\mathrm{B}} = \frac{2}{\sqrt{3}} \sqrt{k_{\mathrm{f}}\left(\bar{\varepsilon}_{\mathrm{pl}}, \dot{\bar{\varepsilon}}_{\mathrm{pl}}, T\right)^2 - 4\,\tau_{yz}{}^2}$$ (12)

Considering the profile cross-section (Figure 7a), stresses and strains can develop partially plastic over the course of the y coordinate (Figure 7b). Additionally, in the partial heated case, assumption 6 must be generalized. Wolter [20] firstly determined that it is possible for the stress- and strain-free fiber to deviate in position and that the position of stress and strain symmetry changes during the bending process. For example, an added normal compressive stress shifts the position of the stress-free fiber in the direction of the compressive area [21]. A similar effect is expected for a partially heated profile. As temperature increases in the heated area, the flow stress is reduced. In the room temperature area, flow stress remains constant. As the force equilibrium needs to be fulfilled, the symmetry axis between the tensile and the compressive zone (the stress-free fiber) needs to shift in direction of the compressive zone. Due to the coupling of stress and strain, the strain free fiber will move in direction of the compressive area too. To solve the integral correlation in Equation (11), it is necessary to calculate the position of the stress-free fiber (y_{m}) and the position at which the material starts plastic forming ($y_{\mathrm{pl,l}}$ in the lower and $y_{\mathrm{pl,u}}$ in the upper area).

The fibers for plasticity onset $y_{\mathrm{pl,l}}$ and $y_{\mathrm{pl,u}}$ at $x \geq l_{\mathrm{p}}$ can be calculated using assumption 9 and Equation (3) through the relation

$$\pm \left(\frac{2(1 - v^2)}{\sqrt{3}E(T)} k_{\mathrm{f}}\left(0, \dot{\bar{\varepsilon}}_{\mathrm{pl}}, T\right) - a_{\mathrm{T}}\Delta T \right) r_{\mathrm{mC}} = y_{\mathrm{pl}}$$ (13)

with the Poisson's ratio v. The shift of the stress-free fiber y_{m} from the center of gravity results from a shift due to normal stresses $y_{\mathrm{m,N}}$ and a shift due to temperature $y_{\mathrm{m,T}}$ yielding in

$$y_m = y_{\mathrm{m,N}} + y_{\mathrm{m,T}}.$$ (14)

The shift due to normal stresses of the stress-free fiber occurs by the superposition of normal stresses [5] as they are added in the Mises stress. The position of the stress-free fiber is temperature dependent because of the reduced flow stress in the heated area. As the flow stress in the room temperature area remains constant, the stress-free fiber has to shift to the compressive zone so that force equilibrium between the tensile and the compressive zone of the cross section is fulfilled. As the normal stresses are negligible (assumption 5), the stress-free fiber position is $y_m = y_{m,T}$.

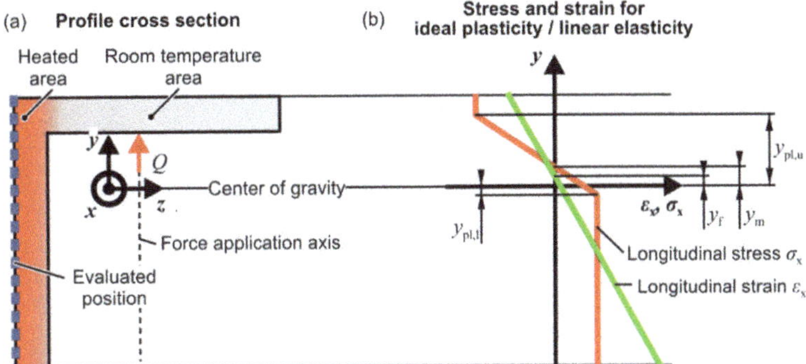

Figure 7. (**a**) profile cross-section of the partially heated profile, (**b**) longitudinal stress and strain progression for ideal plastic and linear elastic material behavior and partially plastic behavior over the cross section.

The position of the stress-free fiber due to temperature $y_{m,T}$ needs to be evaluated to calculate the bending moment. To evaluate $y_{m,T}$, some considerations are necessary. Mechanical equilibrium between tensile and compressive zones for the room temperature case is fulfilled if the stress-free fiber is in the origin of the profile cross section (Figure 8a). Through the lower flow stress, the stress in the heated section is reduced (Figure 8b). As the force equilibrium between both areas still needs to be fulfilled, a shift in stress-free fiber position $y_{m,T}$ is caused. To calculate the position of the stress-free fiber, the forces in the compressive area (F_{RT} and F_h) are related to their corresponding distance at which they act upon. In detail, $\frac{b}{2} - y_0$ is related to the room temperature force F_{RT} and $\frac{b}{2} - y_0 - y_{m,T}$ is related to the force in the partially heated case F_h. y_0 is the distance between the middle of the profile cross section and the center of gravity. Through geometrical considerations, it is then assumed that the share of $y_{m,T}$ on the distance of the compressive zone in the room temperature case ($\frac{b}{2} - y_0$) is the same as the share of the force $F_{RT} - F_h$ on the force at room temperature (F_{RT}) (Figure 8c). This means that the force change between the room temperature case and the partial heated case is related to the stress-free fiber shift.

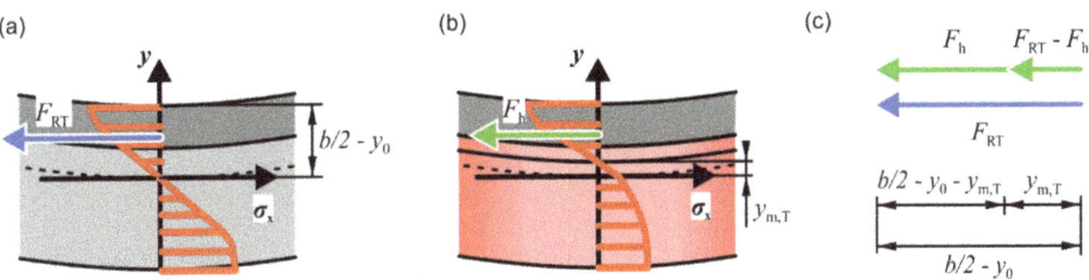

Figure 8. Relation between cross-sectional forces and stress-free fiber. (**a**) Room-temperature case, (**b**) partially heated case, (**c**) Relation between force and stress-free fiber.

The resulting equation for the stress-free fiber can then be expressed as

$$\frac{F_{RT} - F_h}{F_{RT}} = \frac{y_{m,T}}{\frac{b}{2} - y_0}.$$ (15)

Using the integral relation

$$F = \int \sigma_B dA$$ (16)

the forces F_{RT} and F_h can be calculated. With y_m, y_{pl}, and α, the bending moment (Equation (11)) for the process can be calculated (see Appendix F).

4. Analysis of Push-Bending with Partial Cross-Sectional Heating and the Resulting Geometry

The effects of the partial heating strategy are evaluated in this chapter. First, the influence on the product geometry is investigated by examining the unloaded geometry. After that, the position of the plastic zone is determined. Finally, the influence of the partial heating strategy on bending load and warping angle is investigated to understand the mechanisms in the process.

4.1. Analysis of the Unloaded Geometry

The influence of the partial heating on the unloaded bending geometry is shown focusing on springback and warping (Figure 9a) From 25 to 300 °C, the warping angle and springback ratio stay approximately constant. With further increasing temperature springback and warping angle are decreasing. For higher feed velocity the profile springback increases by a mean value of 12% (see Figure 9b). The increase of springback at higher feed velocity can be attributed to the strain rate hardening (Figure 4). The higher feed velocity of 8 mm/s leads to a doubled strain rate compared to the lower velocity of 4 mm/s, resulting in a change of the highest strain rate from 0.055 1/s to 0.11 1/s. This results in an increase of the flow stress by 9% at 0.2 strain for the 600 °C case.

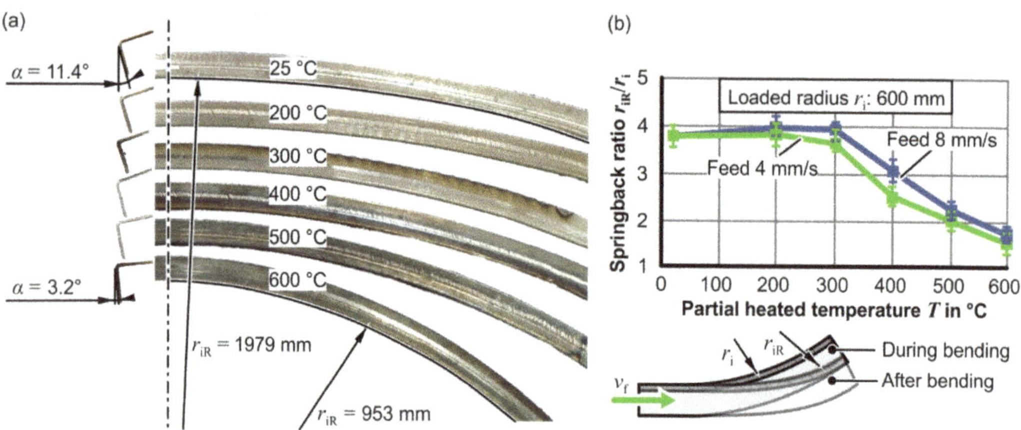

Figure 9. (a) Profile specimens and corresponding cross-sections for a 600 mm loaded bending radius (change of moment arm through the heating area is neglected) and profile feed of 8 mm/s for different temperatures in the sprung back state. (b) springback ratio for varying partial heating temperatures and profile feed velocities.

Until a temperature of 300 °C in the heated area, the maximum change between the room temperature case and the partial heated case is 2%. For temperatures higher than 300 °C, the springback reduces approximately linearly with a maximum reduction of 56% at 600 °C.

To validate the numerical model, the experimental data are compared to the numerical data (Figure 10a). The numerical simulation can approximate the experimental springback with an error of 3%.

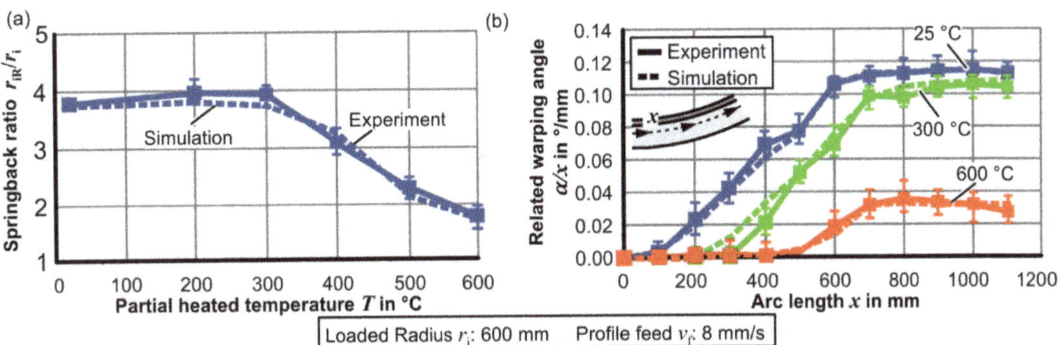

Figure 10. Comparison of springback (**a**) and related warping angle (**b**) between experiments and Simulation for 8 mm/s feed and loaded radius of 600 mm.

The related warping angle increases linearly over the arc length for each parameter set until a saturation value (Figure 10b) on the arc length of 600–800 mm is reached, after saturation, the related warping angle is constant with a deviation from a constant value of maximum 13%. Related warping angle is decreasing with increasing temperature. After saturation ($x = 1100$ mm), the decrease for 300 °C is 9% and for 600 °C the decrease is 76% compared to the room temperature curve. The mean deviation between experimental and numerical results for all parameter sets is 10%. The springback is only reduced at temperatures higher than 300 °C. This can be attributed to the flow stress reduction in the cross section. To further discuss this effect, the position of the principal forming zone must be determined.

4.2. Investigation of the Plastic Zone

To analyze the influence of the partial heating strategy on springback and warping the location of the principal forming zone is important (Figure 11). The actual bending moment in the profile is evaluated using the bending force of the simulation data. The threshold bending moment is received by solving the bending moment equation (Equation (12)) for the minimum flow stress necessary to receive the set loaded radius. Temperature is assumed constant in the heated area. Plasticity starts at the first position the profile bending moment surpasses the threshold bending moment. For the room temperature case (Figure 11a) and 300 °C partial heating temperature (Figure 11b), the plastic deformation limit is surpassed at the position of the counter roll $x = 0$. The flow stress reduction for 300 °C is not sufficient for plasticity to start in the heated area. For the 600 °C case (Figure 11c), the bending moment reduction is sufficient for plasticity to start in the heated area at $l_p = 77$ mm. Consequently, the bending moment is only decreased if the heating temperature is higher than 300 °C. The decreased bending moment results in lowered springback of the profile (Figure 9).

As plasticity for the 600 °C case does not start at the position of the counter roll, the moment arm also decreases which influences the set bending radius. The loaded radius is not adjusted for this effect in the analysis as it was not possible to determine the true radius before the analysis. The set radius r_i for the 600 °C case would then be 447 mm, which would cause a deviation in the springback ratio of 11%. Consequently, the change of springback ratio between the 600 °C case and the room temperature case would be 44% (Figure 9b).

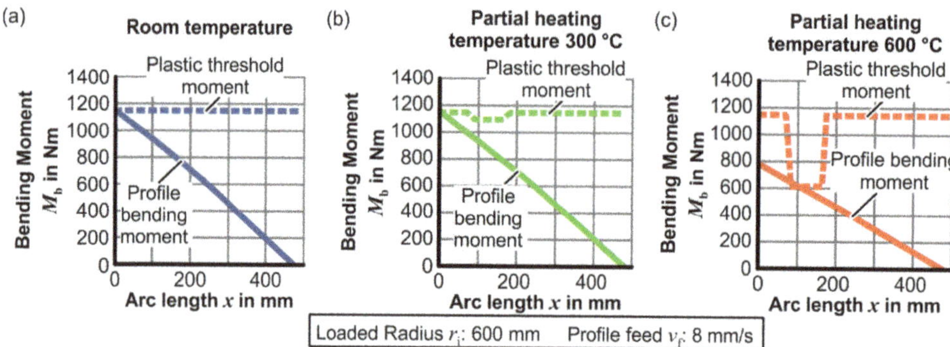

Figure 11. Bending moments and plastic threshold bending moments for room temperature (**a**), 300 °C partial heating temperature (**b**) and 600 °C partial heating temperature (**c**) case resulting from mechanical equilibrium and flow stress for the loaded radius.

4.3. Analysis of the Bending Load and Profile Warping in the Kinematic Push Bending Phase

The longitudinal stresses σ_x in the profile cross-section are displayed to examine the influence of the stress-free fiber position on the profile warping and to investigate the accuracy of the analytical model. (Figure 12). While the stress for the room temperature case and the 300 °C cases behave elastic-plastic, the stress for the 600 °C case is fully plastic. Compared to the room temperature case, the maximum stress in the tensile area is reduced by 4% for the 300 °C partial heating temperature and by 52% for 600 °C partial heating. The position of the stress-free fiber y_m related to the profile width b is 0.013 for the room temperature case, 0.057 for 300 °C heating temperature, and 0.14 for 600 °C heating temperature. The warping reduction can be traced back to the change in stress free fiber position. The shift of stress-free fiber for 300 °C explains why warping is reduced (Figure 10b), but the bending moment is the same as for the room temperature case (Figure 11a,b). Through partial heating, stresses in the tensile area become lower. As force equilibrium between tensile and compressive area still needs to be fulfilled the stress-free fiber shifts in direction of the room temperature area. Through the flow stress reduction shear stresses also get reduced in the heated area. Consequently, with increasing stress-free fiber position, the torsion moment in the profile is reduced. As the increase in stress free fiber position is higher for 600 °C than for 300 °C, the warping reduction is higher. The mean deviation between experimental and analytical data is 7%. The model can therefore predict the stress curves in the profile cross-section and the shift of the stress-free fiber.

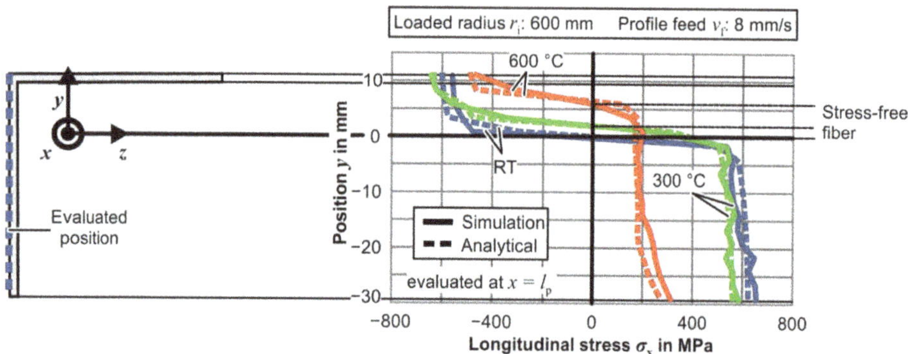

Figure 12. Comparison of numerical and analytical results for longitudinal stress over the profile y-axis at the onset of plastic deformation for the loaded radius of 600 mm and profile feed of 8 mm/s for different partial heating temperatures.

To investigate the accuracy at which the analytical model can predict the bending moment and the profile warping as well as to investigate the warping in the loaded state, bending moment and related warping angle are displayed (Figure 13). Both are evaluated at the start of the plastic zone ($x = l_p$). For the analysis of the related warping angle, the point of zero warping for the 600 °C the curve is shifted to the origin of the diagram to allow an easier comparison with the room temperature and 300 °C cases. The bending moment (Figure 13a) decreases linearly until the position of the bending roll ($x = 480$ mm) is reached. The bending moment is the highest for room temperature, though the mean difference between room temperature case and 300 °C case is 1%. The bending moments for 300 °C and room temperature cases are the same because they both start plastic deformation at the same threshold bending moment (Figure 11a,b). Compared to the highest bending moment, the highest bending moment for the 600 °C case decreases by 40%. In all cases, the analytical moment overestimates the numerical bending moment with a mean deviation of 7%. The reason for the deviation could be the neglected influence of the normal stresses (assumption 5) or a cumulative effect of the error margins resulting from both numerical and analytical analysis compared to the experimental data. The related warping angles (Figure 13b) are only evaluated at the positions between counter roll and bending roll as the analytical model is only applicable for the loaded state. Thus, the data displayed here are not directly comparable to the unloaded warping angle (Figure 10b). The related warping angles increase approximately linearly as they result from the equation for torsion moment (Equation (10)). The torsion moment behaves linearly over the x-axis. The maximum related warping angle of the 300 °C case is decreased by 13% and the maximum related warping angle for the 600 °C case is decreased by 53% compared to the room temperature case. This is analogical to the unloaded state. The analytical model can predict the numerical data with a mean deviation of 10%. It is thus possible for the analytical model to predict the warping and bending moment. The analytical model can now be used for process development and to generalize the results of this study for other profile geometries and materials.

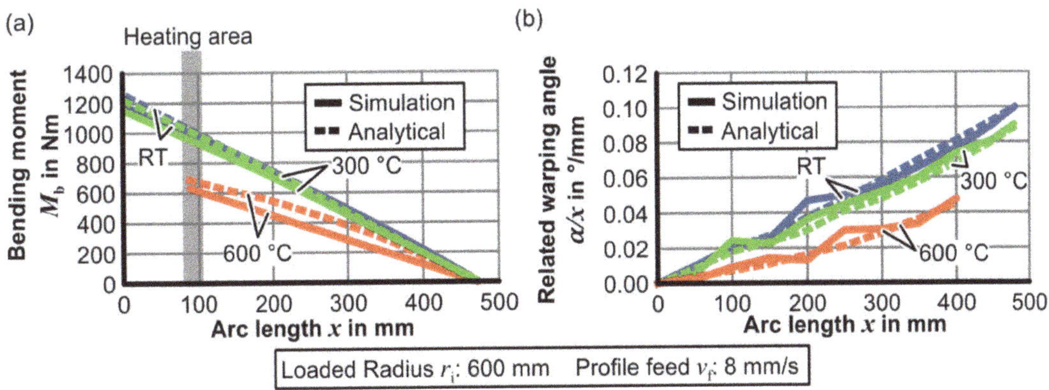

Figure 13. Analytical and numerical bending moment (**a**) and related waring angle (**b**) for the continuous push-bending phase at room temperature, 300 °C and 600 °C for a loaded radius of 600 mm and profile feed of 8 mm/s as a function of profile arc length.

5. Conclusions

To reduce warping and springback in the bending of profiles with asymmetric geometry in the force application axis, partial heating of the cross-section can be used. It has been shown that partial heating of the cross section leads to a springback reduction of at least 44% and a warping reduction of 76% compared to the room temperature case. The warping and springback reduction can be attributed to a shift in the stress-free fiber position. Partial heating reduces the flow stress in the heated area. Through the flow stress reduction, a shift in stress-free fiber to the compressive zone is noticeable, reducing the stresses in the cross section. To achieve reduced springback and warping, a threshold temperature of 300 °C must be reached.

Additionally, an analytical model has been developed which is able to predict warping with 90% and bending moment with 93% accuracy in a kinematic push-bending process. This model can be used for process design and control.

Author Contributions: Conceptualization, R.M. and A.E.T.; methodology, E.H.; formal analysis, E.H.; investigation, E.H.; resources, A.E.T.; writing-original draft preparation, E.H.; writing-review and editing, R.M. and A.E.T.; visualization E.H.; supervision, R.M. and A.E.T.; project administration, E.H. and R.M.; funding acquisition, A.E.T. All authors have read and agreed to the published version of the manuscript.

Funding: The authors would like to thank the German Research Foundation DFG for the support of the depicted research through project no. 408302329. The financial support is greatly acknowledged.

Institutional Review Board Statement: Not applicable.

Informed Consent Statement: Not applicable.

Data Availability Statement: Data sharing is not applicable to this article.

Acknowledgments: The authors thank the German Research Foundation (DFG) for the financial support of project 408302329 "Kinematic profile bending with locally heated cross-section" (German: Kinematisches Profilbiegen mit partieller Erwärmung des Querschnitts). The authors would also like to thank Till Clausmeyer for his helpful comments on the manuscript. His help is greatly appreciated.

Conflicts of Interest: The authors declare no conflict of Interests.

Appendix A

The parameters for flow curves are fitted through the least square method using the experimental data (Table A1).

Table A1. Flow curve extrapolation parameters.

Temperature	C_1	C_2	C_3	C_4	C_5	C_6	C_7	C_8	$\dot{\varepsilon}_{pl,0}$
25 °C	562	445	28	0	0	0	0	0	0.0003
300 °C	541	440	37	0.6	8	0.09	0	0	0.0003
400 °C	470	393	44	10	13	0.1	178	0.55	0.0003
500 °C	325	285	65	7.3	5	0.2	8.4	0	0.0003
600 °C	177	160	367	12.9	8	0.2	0	4.8	0.0003

Appendix B

To validate the analytical assumption regarding constant temperature in heated area and area at room temperature in the kinematic push-bending phase, the temperature fields in the simulation and experiment are evaluated (Figure A1). The thermographic measurements were carried out using a VarioCam HD produced by InfraTec, Dresden, Germany. For both the numerical and experimental temperature distribution the temperature near, the area at room temperature cools down to up to 400 °C. As the forming zone is at the beginning of the heated zone, the difference of temperature distribution between numerical and analytical results is negligible.

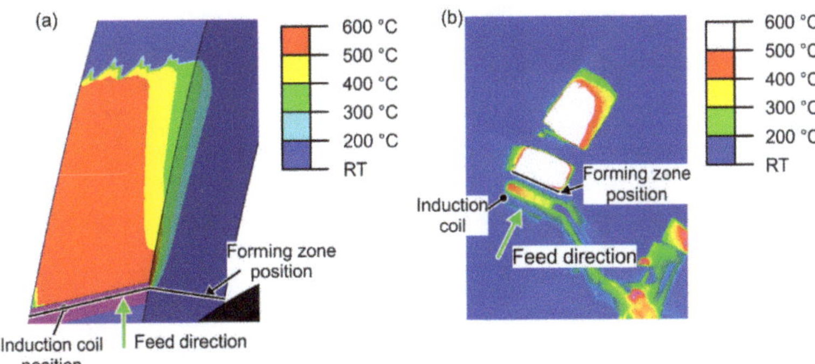

Figure A1. Temperature distribution in the kinematic push-bending phase for 600 °C heating temperature and 8 mm/s feed velocity. (**a**) simulated temperature distribution, (**b**) thermographic measurement.

Appendix C

For the validation of the neglected elastic and thermal strains in the plastic region for the analytical model, the elastic and total strain resulting from the numerical simulation are evaluated (Figure A2). For room temperature, the error resulting from neglecting the elastic strains is 6%. From 200 °C, each increase of 100 °C increases this error by approximately 1% as thermal strain increases. The maximum error from excluding thermal and elastic strains occurs at 600 °C partial heating temperature. The resulting error for the strains is 10%.

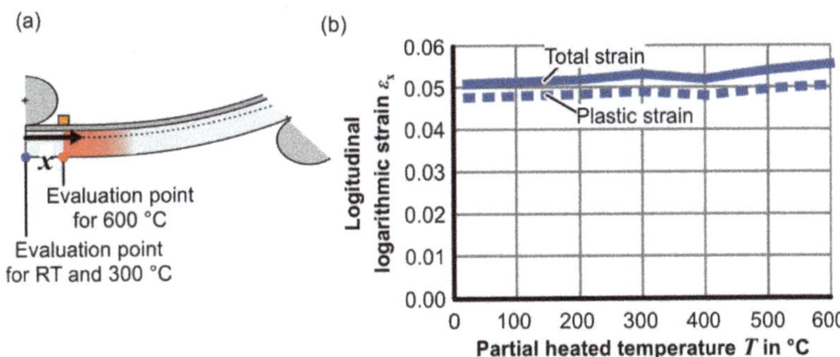

Figure A2. Difference between plastic strain and total strain. (**a**) evaluated position, (**b**) total strain and plastic strain evaluated through FE simulation for different partial heating temperatures.

Appendix D

The slope of a curve $f(x)$ at a position x is defined using the relations

$$m = \frac{df(x)}{dx}. \tag{A1}$$

As the slope of $f(x)$ can be dependent on the position x a mean value of the slope can be calculated with

$$\overline{m} = \frac{\int_x m\,dx}{\Delta x}. \tag{A2}$$

Plastic strain and strain rates are dependent on the y-axis position in the cross-section. Consequently, to calculate a mean slope of the flow curve, a mean value over strains and strain rates is necessary to approximate the hardening behavior. Using Equations (A1) and (A2), the mean slope of the flow curve over the strain rate $E_{\mathrm{p,r}}\left(\bar{\varepsilon}_{\mathrm{pl}}, T\right)$ can be calculated to:

$$E_{\mathrm{p,r}}\left(\bar{\varepsilon}_{\mathrm{pl}}, T\right) = \frac{\int_{\dot{\bar{\varepsilon}}_{\mathrm{pl}}} \frac{d}{d\dot{\bar{\varepsilon}}_{\mathrm{pl}}} k_{\mathrm{f}}(\bar{\varepsilon}_{\mathrm{pl}}, \dot{\bar{\varepsilon}}_{\mathrm{pl}}, T) d\dot{\bar{\varepsilon}}_{\mathrm{pl}}}{\Delta \dot{\bar{\varepsilon}}_{\mathrm{pl}}}. \tag{A3}$$

Using Equation (A3), the mean slope of the flow curve over strain and strain rate is

$$E_{\mathrm{p}}(T) = \frac{\int_{\bar{\varepsilon}_{\mathrm{pl}}} \left(\frac{d}{d\bar{\varepsilon}_{\mathrm{pl}}} E_{\mathrm{p,r}}\left(\bar{\varepsilon}_{\mathrm{pl}}, T\right) \right) d\bar{\varepsilon}_{\mathrm{pl}}}{\Delta \bar{\varepsilon}_{\mathrm{pl}}}. \tag{A4}$$

With the mean slope of the flow curve, the relation between shear stresses and shear strain rate can be approximated.

Appendix E

The integral for the torsion moment M_{T} for the profile geometry reads

$$M_{\mathrm{T}} = \int_{-\left(\frac{b}{2}-z_0-t\right)}^{\frac{b}{2}+z_0} \int_{\frac{b}{2}-y_0-t}^{\frac{b}{2}-y_0} \tau_{yz,\mathrm{RT}} t \, dydz + \int_{-\left(\frac{b}{2}-z_0\right)}^{\left(\frac{b}{2}-z_0-t\right)} \int_{-\left(\frac{b}{2}+y_0\right)}^{\frac{b}{2}-y_0} \tau_{yz,\mathrm{h}} t \, dydz \tag{A5}$$

where z_0 and y_0 are the correspondent coordinate distances to the profile center to the center of gravity, $\tau_{yz,\mathrm{RT}}$ is the shear stress in the room temperature area, and $\tau_{yz,\mathrm{h}}$ is the shear stress at the heated area. Evaluation of Equation (9) yields

$$\tau_{yz,\mathrm{RT}} = \frac{Q \cdot \left(b - \frac{b^2 t + (b-t)t^2}{2(bt+(b-t)t)} - z\right) \left(\frac{b^2 t + (b-t)t^2}{2(bt+(b-t)t)} - \frac{t}{2}\right)}{\frac{tb^3}{3} + \frac{(b-t)t^3}{3} - \frac{(b^2 t + (b-t)t^2)^2}{4(bt+(b-t)t)}} \tag{A6}$$

$$\tau_{yz,\mathrm{h}} = -\frac{Q \cdot \left(b - \frac{b^2 t + (b-t)t^2}{2(bt+(b-t)t)} + y\right) \left(y - \frac{t}{2}\right)}{\frac{tb^3}{3} + \frac{(b-t)t^3}{3} - \frac{(b^2 t + (b-t)t^2)^2}{4(bt+(b-t)t)}}. \tag{A7}$$

for the corresponding profile geometry. Solving the torsion moment (Equation (A5)) using both shear stress from force equilibrium $M_{\mathrm{T,e}}$ (Equation (A6)) and shear stress from strains $M_{\mathrm{T,s}}$ (Equation (10)), the warping angle α can be calculated numerically using the relation

$$M_{\mathrm{T,e}} = M_{\mathrm{T,s}}. \tag{A8}$$

Appendix F

The bending moment $M_{\mathrm{B,pp}}$ for RT and 300 °C is given by

$$\begin{aligned}
M_{\mathrm{B,pp}} = \; &\int_{-\left(\frac{b}{2}-z_0-t\right)}^{\frac{b}{2}+z_0} \int_{\frac{b}{2}-y_0-t}^{\frac{b}{2}-y_0} \sigma_{\mathrm{B,pl}}(T = RT) y \, dydz \\
&+ \int_{-\left(\frac{b}{2}-z_0\right)}^{\left(\frac{b}{2}-z_0-t\right)} \int_{y_{\mathrm{pl,u}}}^{\frac{b}{2}-y_0} \sigma_{\mathrm{B,pl}}(T = T_1) y \, dydz \\
&+ \int_{-\left(\frac{b}{2}-z_0\right)}^{\left(\frac{b}{2}-z_0-t\right)} \int_{y_{\mathrm{m}}}^{y_{\mathrm{pl,u}}} \sigma_{\mathrm{B,el}}(T = T_1) y \, dydz \\
&- \int_{-\left(\frac{b}{2}-z_0\right)}^{\left(\frac{b}{2}-z_0-t\right)} \int_{y_{\mathrm{pl,l}}}^{y_{\mathrm{m}}} \sigma_{\mathrm{B,el}}(T = T_1) y \, dydz \\
&- \int_{-\left(\frac{b}{2}-z_0\right)}^{\left(\frac{b}{2}-z_0-t\right)} \int_{-\left(\frac{b}{2}+y_0\right)}^{y_{\mathrm{pl,l}}} \sigma_{\mathrm{B,pl}}(T = T_1) y \, dydz,
\end{aligned} \tag{A9}$$

with elastic bending stress $\sigma_{B,el}$, plastic bending stress $\sigma_{B,pl}$, the upper plasticization fiber position $y_{pl,u}$, and the lower plastification fiber position $y_{pl,l}$ assuming partial cross-sectional plasticity. For the 600 °C case with full plastification, the integral is expressed as

$$
\begin{aligned}
M_{B,fp} = &\int_{-(\frac{b}{2}-z_0-t)}^{\frac{b}{2}+z_0} \int_{\frac{b}{2}-y_0-t}^{\frac{b}{2}-y_0} \sigma_{B,pl}(T=RT)y\,dy\,dz \\
&+ \int_{-(\frac{b}{2}-z_0)}^{(\frac{b}{2}-z_0-t)} \int_{y_m}^{\frac{b}{2}-y_0} \sigma_{B,pl}(T=T_1)y\,dy\,dz \\
&- \int_{-(\frac{b}{2}-z_0)}^{(\frac{b}{2}-z_0-t)} \int_{-(\frac{b}{2}+y_0)}^{y_m} \sigma_{B,pl}(T=T_1)y\,dy\,dz.
\end{aligned}
\tag{A10}
$$

Using the first degree Taylor approximation on the integrals, the bending moment can be solved.

References

1. Singh, D.K. *Strength of Materials*; Springer International Publishing: New York, NY, USA, 2021; ISBN 978-3-030-59666-8.
2. Seely, F.B.; Putnam, W.J.; Schwalbe, W.L. The Torsional Effect of Transverse Bending Loads on Channel Beams. In *University of Illionois Bulletin 211*; University of Illinois: Champaign, IL, USA, 1930.
3. Vollertsen, F.; Sprenger, A.; Kraus, J.; Arnet, H. Extrusion, channel, and profile bending: A review. *J. Mater. Process. Technol.* **1999**, *87*, 1–27. [CrossRef]
4. Chatti, S.; Hermes, M.; Tekkaya, A.E.; Kleiner, M. The new TSS bending process: 3D bending of profiles with arbitrary cross-sections. *CIRP Ann.* **2010**, *59*, 315–318. [CrossRef]
5. Chatti, S. Optimierung der Fertigungsgenauigkeit beim Profilbiegen. Ph.D. Thesis, Universität Dortmund, Dortmund, Germany, 1998.
6. Kreye, B. Verfahren und Vorrichtung zum Biegen von hohlen Metallprofilen, insbesondere Rahmenprofilen für Fenster und Türen. Patent EP0115840, 15 August 1984.
7. Li, P.; Wang, L.; Li, M. Flexible-bending of profiles with asymmetric cross-section and elimination of side bending defect. *Int. J. Adv. Manuf. Technol.* **2016**, *87*, 2853–2859. [CrossRef]
8. Wang, A.; Xue, H.; Bayraktar, E.; Yang, Y.; Saud, S.; Chen, P. Analysis and control of twist defects of aluminum profiles with large z-section in roll bending process. *Metals* **2020**, *10*, 31. [CrossRef]
9. Jahn, E. Herstellung von Rohrbögen nach dem Induktivbiegeverfahren. *Tech. Überwach.* **1976**, *17*, 221–227.
10. Park, J.C.; Seong, D.Y.; Yang, D.Y.; Cha, M.H. Development of an innovative bending process employing synchronous incremental heating and incremental forming. In Proceedings of the 10th International Conference on Technology of Plasticity, ICTP 2011, Aachen, Germany, 25–30 September 2011; pp. 325–330.
11. Behne, T. A New Bending Technique for Large-Diameter Pipe. *Iron Age Metalwork. Int. IAMI* **1983**, *22*, 34–36.
12. Yanagimoto, J.; Oyamada, K. Mechanism of springback-free bending of high-strength steel sheets under warm forming conditions. *CIRP Ann. Manuf. Technol.* **2007**, *56*, 265–268. [CrossRef]
13. Avent, R.R. *Guide for Heat-Straightening of Damaged Steel Bridge Members*; Federal Highway Administration: Washington, DC, USA, 2008.
14. Rothman, R.L. *Flame Straightening Quenched-and-Tempered Steels in Ship Construction*; Technical Report for US Ship Structure Comitee: Columbus, OH, USA, November 1975.
15. Kraus, J. Laserstrahlumformen von Profilen. Ph.D. Thesis, Friedrich-Alexander-Universität Erlangen-Nürnberg, Erlangen, Germany, 1997.
16. Li, W.; Yao, Y.L. Laser bending of tubes: Mechanism, analysis, and prediction. *J. Manuf. Sci. Eng. Trans. ASME* **2001**, *123*, 674–681. [CrossRef]
17. Raither, W.; Bergamini, A.; Ermanni, P. Profile beams with adaptive bending–twist coupling by adjustable shear centre location. *J. Intell. Mater. Syst. Struct.* **2013**, *24*, 334–346. [CrossRef]
18. Ludwik, P. Technologische Studie über Blechbiegung—Ein Beitrag zur Mechanik der Formänderungen. *Tech. Blätter* **1903**, *35*, 133–159.
19. Haupt, P. *Continuum Mechanics and Theory of Materials*; Springer: Berlin, Germany, 2002; ISBN 978-3-642-07718-0.
20. Wolter, K. *Freies Biegen von Blechen*; Dt. Ingenieur-Verl: Düsseldorf, Germany, 1952.
21. Timoshenko, S.P.; Gere, J.M. *Theory of Elastic Stability*; McGraw-Hill: New York, NY, USA, 1961.

Article

Towards Manufacturing of Ultrafine-Laminated Structures in Metallic Tubes by Accumulative Extrusion Bonding

Matthew R. Standley and Marko Knezevic *

Department of Mechanical Engineering, University of New Hampshire, Durham, NH 03824, USA;
mrx46@wildcats.unh.edu
* Correspondence: marko.knezevic@unh.edu

Abstract: A severe plastic deformation process, termed accumulative extrusion bonding (AEB), is conceived to steady-state bond metals in the form of multilayered tubes. It is shown that AEB can facilitate bonding of metals in their solid-state, like the process of accumulative roll bonding (ARB). The AEB steps involve iterative extrusion, cutting, expanding, restacking, and annealing. As the process is iterated, the laminated structure layer thicknesses decrease within the tube wall, while the tube wall thickness and outer diameter remain constant. Multilayered bimetallic tubes with approximately 2 mm wall thickness and 25.25 mm outer diameter of copper-aluminum are produced at 52% radial strain per extrusion pass to contain eight layers. Furthermore, tubes of copper-copper are produced at 52% and 68% strain to contain two layers. The amount of bonding at the metal-to-metal interfaces and grain structure are measured using optical microscopy. After detailed examination, only the copper-copper bimetal deformed to 68% strain is found bonded. The yield strength of the copper-copper tube extruded at 68% improves from 83 MPa to 481 MPa; a 480% increase. Surface preparation, as described by the thin film theory, and the amount of deformation imposed per extrusion pass are identified and discussed as key contributors to enact successful metal-to-metal bonding at the interface. Unlike in ARB, bonding in AEB does not occur at ~50% strain revealing the significant role of more complex geometry of tubes relative to sheets in solid-state bonding.

Keywords: plasticity; strength; metallic tubes; finite element analysis; accumulative extrusion bonding

Citation: Standley, M.R.; Knezevic, M. Towards Manufacturing of Ultrafine-Laminated Structures in Metallic Tubes by Accumulative Extrusion Bonding. *Metals* **2021**, *11*, 389. https://doi.org/10.3390/met11030389

Academic Editors: Andrew Kennedy, Gabriel Centeno Báez and Maria Beatriz Silva

Received: 15 January 2021
Accepted: 23 February 2021
Published: 27 February 2021

Publisher's Note: MDPI stays neutral with regard to jurisdictional claims in published maps and institutional affiliations.

1. Introduction

Bimetallic materials have been used for components delivering different material properties by their geometry (e.g., inside versus outside of a tube) [1–4]. Such components achieve benefits, such as lower cost for the consumer and producer, reduced weight, simplification of design, and/or reduced number of parts in a structure or assembly [3]. Bimetallic materials could be manufactured in the form of tubular geometries to serve desired applications. When employing bimetallic tubes, for example, one material can provide strength and stability, while the other can offer better corrosion resistance. A two-layer copper-steel tube, for example, can handle high loads via the steel and the corrosion resistance via the copper [4].

Bonding between the bimetals is not always necessary, and ultimately depends on the application. The two-layer designs typically rely on each material to perform one aspect of the intended design function independent of the other. In this case, a very tight compression fit between the two constituent metals may be appropriate. Evolving from the two-layer concept, multilayered material is envisioned for even more demanding or unique applications. To this end, multilayered materials provide a blended, and most often optimized, set of material properties but require each layer to be bonded to the next to do so.

Multilayered bimetallic manufacturing is a relatively new frontier in manufacturing and delivers superior material characteristics when compared to the constituent materials.

Many material combinations have been reported as bonded using accumulative roll bonding (ARB) such as Cu/Ti [5], Al/Cu [6], Al/Zn [7], Mg/Al [8], Cu/Zn/Al [9], Cu/Zn [10], Zr/Nb [11,12], Mg/Nb [13,14] and Zn/Sn [15] in plate form. When the layering is pushed to the ultrafine micron, or ultimately nanometer scale, the multilayered bimetallic material exhibits significantly improved strength [16–24], thermal stability [25,26], resistance to shock damage [27], and resistance to radiation damage [28,29]. Beyond this, the authors of [30] summarize the history of laminated metal composites and other benefits of bimetallic materials, such as improved fracture resistance, delayed fatigue crack growth, and ballistic energy absorption.

This work explores a processing methodology for manufacturing multilayered bimetallic tubing to achieve similar improvements in material properties to ARB sheets. To the knowledge of the authors at the time of publication, no research has reported producing multilayered bimetallic tubing using any severe plastic deformation processes. Following previous research in extrusion to achieve bimetallic tubing [31], this work takes inspiration from their design to develop a more complete manufacturing process. The process is termed accumulative extrusion bonding (AEB) and is used, in an iterative sense, to create single metal and bimetallic tubing with several layers. AEB, like ARB, is a severe plastic deformation process, which is defined as a metal forming process that creates very high strain without significant change to the overall dimensions to produce substantial grain refinement using severe straining and high levels of hydrostatic pressure [11,32–39]. By doing so, a decrease in grain size can improve the material properties following the Hall-Petch relationship [40–42] and increase the material strength by a factor of three to eight [43]. Additionally, as reported in [44], materials with ultrafine grains can have good damping properties, exhibit lower temperature super-plasticity, and high magnetic properties.

The AEB process involves iterative extrusion, cutting, expanding, restacking, and annealing. Due to the increased complexity of maintaining the geometrical shape of tubing relative to sheets, AEB is a much more challenging process than ARB. Other severe plastic deformations processes exist to produce extruded tube, and are well documented in the following review article [45]. Such processes include Equal Channel Angular Pressing (ECAP), Tube Channel Pressing (TCP), Tubular Channel Angular Pressing (TCAP), among others [45–47]. Since restacking and reprocessing is essential in producing ultrafine grains in multilayered tubing, AEB is described herein.

When this AEB process is compared to other AEB processes, the main differences include geometry of the extruded material, the custom dies and setup used, and the expansion process. Additionally, AEB performed in [48] did not use intermittent annealing, and because of this, tracked the true strain increase as samples were continuously processed. The AEB process used in the present research, utilizes annealing after every severe plastic deformation step such that accounting for continuous strain is not needed as strain effects are removed. Additionally, no material specimen or die preheat is used as indicated in the research performed in [49]. Most unique, no research has reported using AEB to manufacture multilayered bimetals in the form of tube as current research utilizes plate or sheet as done in [48–51]. For this reason, as part of the AEB process used for this research, expansion is necessary to facilitate restacking.

Two sets of specialty tooling are designed for a hydraulic press: One imparting 52% radial strain and another imparting 68% radial strain to produce the tubing. The design process is aided by finite element (FE) simulations to better understand the mechanics of the severe plastic deformation operations. After making the tooling, several similar and dissimilar metallic tubes are created to evaluate the extent of bonding, microstructure, and material properties. Hardness of the materials, yield strength, and ultimate tensile strength are compared before and after processing. Moreover, grain structure and the amount of bonding at the metal-to-metal interfaces are measured using optical microscopy. Surface preparation and the amount of deformation imposed per extrusion pass are discussed as critical to successful metal-to-metal bonding at the interface. Future work will attempt to create ultrafine multilayered bimetallic tubes with micron to nano radial layers of alternating material with many metal-metal interfaces governing a unique set of material properties.

2. Methods

2.1. Theory of Bonding

The thin film theory prevails as the primary explanation for bimetal bonding in high pressure cold rolling. Like rolling, extrusion is also a high-pressure process in which the theory is viable as other mechanisms that explain bonding are unlikely to occur. Other such bonding mechanisms are diffusion, overcoming energy barrier, and joint recrystalliza-tion [52]. A brief summary of the theory is as follows:

1. A very thin brittle surface must exist on both metallic faces to be bonded.
2. Under high pressure, the metallic faces are forced into one interface where the thin brittle surfaces on both metals begin to crack under a significant amount of im-posed strain.
3. Through the small cracks fresh virgin material extrudes which interact with the opposing virgin material to form a metallic bond.

In ARB applications, the brittle surfaces are prepared by light scratch brushing using stainless brushes, and the high pressure is provided by rolls. For example, nickel plated Cu/Al [53], Mg/Nb [13], Al/Ni [54], and Al/Al [55] sheets were produced using this technique. In this work, high pressure is provided by a die, a mandrel, and a punch mounted in a hydraulic press. The theory and extrusion process are depicted schematically in Figure 1. In stage 1, as indicated by the bubble numbers, the two metal tubes are under compression due to the punch (not shown), and an initial air gap is present at the metal-metal interface. This surface must be as clean as possible, free of any contamination, and prepared such that a thin brittle surface exists. As the material is forced into the extrusion ledge in stage 2, the air gap is significantly reduced, the metal-to-metal interface is formed, and plastic strain occurs within the metals. Due to the high strain levels, cracks form within the thin brittle surfaces, and virgin material of each metal extrudes through the cracks, interact, and bond. Entering stage 3 completes the extrusion process by providing the final desired shape: a reduced outer diameter and maintained inner diameter of a new tube size. Some areas may not bond, and voids may become present. Further processing, by repeating the process shown in Figure 2, will continue to thin and stretch the interfaces such that voids, and trapped oxides will be thinned and blended into the metallic structure where their influence on material behavior is minimized [13].

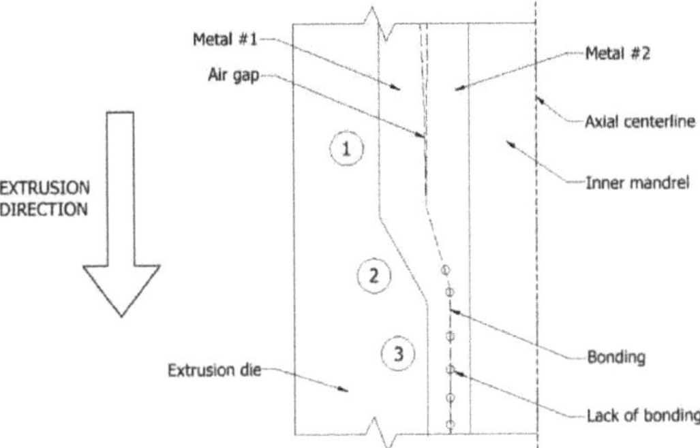

Figure 1. Axis-symmetric cross section of extrusion process and the three stages of bonding using AEB. The metal initially experiences compression (stage 1) before entering the extrusion zone (stage 2), where severe plastic deformation occurs, and exits in the final desired shape (stage 3).

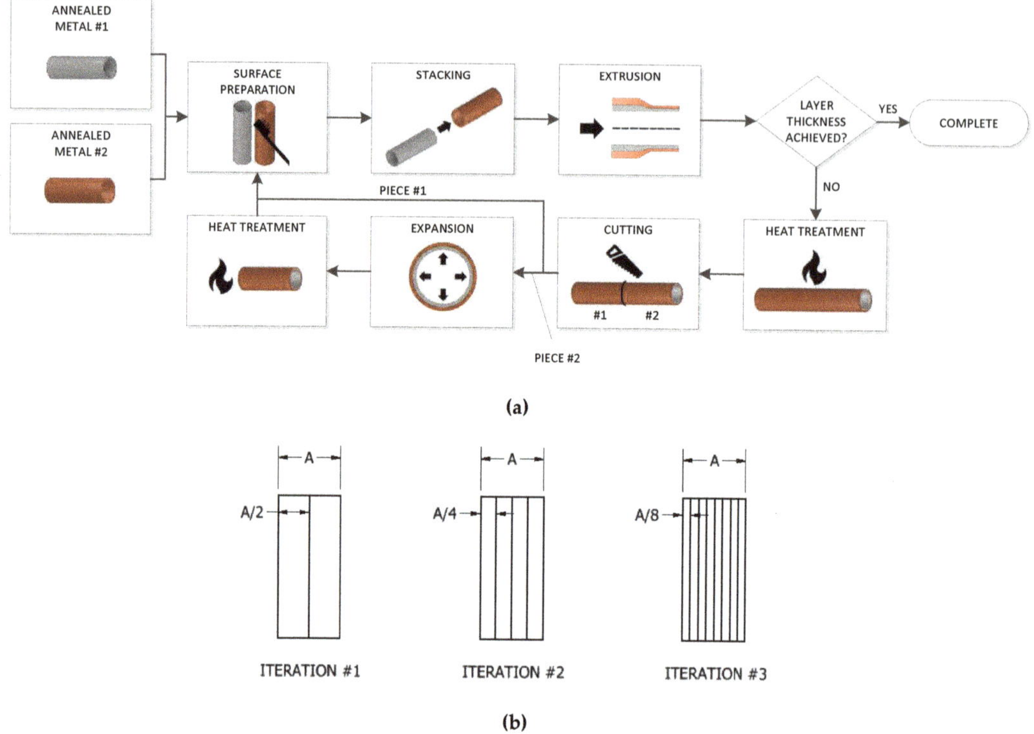

Figure 2. (**a**) The manufacturing process flow map for achieving ultrafine-laminated structures in metallic tubes via AEB. (**b**) Graphical visual of the nominal layer thickness within the tube wall when a bimetal is processed multiple times using AEB. When processing using AEB, the wall thickness is maintained while individual layers exponentially decrease. Note that processing steps, such as cleaning and expansion, will impact layer thicknesses and overall wall thickness such that each individual layer is not expected to be exactly the same.

2.2. Manufacturing

To achieve multilayered bimetallic tubing, it is essential to introduce an iterative process to obtain layers expediently. As the process is iterated, the layer thicknesses within the tubular wall exponentially decreases, while the number of layers exponentially increases using the process flow detailed in Figure 2. Unlike the research in [56], which manufactured bimetallic tube from solid billets of copper and aluminum, the starting base material is tube since tubes are ubiquitous and can be purchased such that they initially stack together.

The first process step is surface preparation of initially annealed tubes followed by stacking. Surface preparation is important to remove foreign material and naturally occurring oxides which can hinder bonding. Additionally, during this step, the surfaces to be bonded must be hardened as described by the thin film theory. Once prepared, one metal tube is inserted into the other metal tube and then the stacked tube is extruded by decreasing the outer diameter and maintaining the inner.

If the desired laminated layer thickness is not achieved, then the metals are prepared for another iteration as shown in Figure 2. The first step to prepare the extruded tube for reprocessing is annealing to restore ductility and bisection at the midpoint to create two tubes of approximately half the extruded length. One of the two metal tubes is then expanded such that it can fit over its extruded diameter. After another annealing of the expanded tube, the initial process is repeated. This continues until the desired layer thickness is achieved. The exponential decrease in layer thickness, occurring at 2^i, is critical in achieving very thin layer thicknesses in a reasonable way.

The two annealing steps, and the initial annealing, are critical to processing and are tactically used to restore ductility before all severe plastic deformation process steps. Based on testing performed, herein, the process flow shown in Figure 2 is the minimum process flow required. Bimetallic tubes produced when omitting any of the annealing steps, for example, caused blistering and tearing during extrusion or expansion.

2.2.1. Surface Preparation

The stacking operation is the simple action of inserting one metal tube into the other. Before this operation is performed, it is critically important to prepare the surfaces, which will become the metal-to-metal interface. In ARB, the interfacial surfaces are typically degreased using acetone and then scratch-brushed with stainless steel bristles [13,53–55]. This surface preparation is reportedly one of the most important steps to achieve full bonding, since it removes the naturally occurring oxide layer and hardens the surface simultaneously. The scratch brushing creates a slightly hardened and brittle outer surface in comparison to the bulk material, due to local strain hardening occurring at the surface. The brittle surface, which will be prone to cracking during extrusion, will allow virgin sub-surface metal to pass through the cracks to contact the virgin material of the other metal to enact bonding. Scratch brushing is applied transverse to the extrusion direction to help promote crack opening.

Before stacking and scratch brushing, acid cleaning is performed to remove any surface impurities. The copper was cleaned by pickling using a solution of 10% sulfuric acid and 90% distilled water per volume. This was done at room temperature for 10 min. After, the acid was neutralized with cool distilled water. The copper was then degreased in an ultrasonic acetone bath for 30 min, where the acetone was drained and replenished halfway through the cleaning process. The aluminum only received degreasing using the ultrasonic acetone bath. After degreasing in acetone, the metal tubes were scratch-brushed and stacked together.

To achieve the most optimal hardened surface, tubes were scratched with stainless steel bristles. The outer diametrical interface was scratched with a handheld stainless-steel brush, while the inner diametrical interface was brushed with a rotary stainless-steel brush. The stainless-steel bristles were 25.4 mm long on the handheld brush with a diameter of 0.305 mm. The stainless-steel bristles on the rotary brush were 13.97 mm long with a diameter of 0.152 mm and rotated at a constant RPM during application. The two different methods of applications were employed due to the curved geometry of each surface.

The surface roughness was measured before and after brushing and is tabulated in Table 1. Surface roughness increased 29.6% and 46.8% respectively on the inner, and outer surfaces, respectively. To find the average surface finish, measurements were taken in 10 random locations on each surface using a Pocket Surf III profilometer (Mahr Federal Inc., Providence, RI, USA). All surface finish values reported before and after brushing are representative of cold extruded aluminum using lubricant for context [57].

Table 1. Average surface roughness before and after scratch brushing.

Process	Surface	Material	Before μm RA√	After μm RA√	% Increase
Rotary brush	Inner diameter	Copper	1.15	1.49	29.6
Handheld brush	Outer diameter	Aluminum	0.32	0.47	46.8

The intent of the brushing is not to induce visible asperities and significantly increase the surface roughness, but to clean the surface of oxidation while hardening it at the same time. Oxidation layer minimization is necessary to aid in bonding, but significantly over-brushing did not improve the amount of bonding. Not quantified by the authors, but reported in literature [52], minimizing the build-up of oxidation is critical to aid in bonding and to reduce additional foreign material inclusion. Beyond scratch brushing, minimizing contact with the atmosphere was employed to discourage further oxidation growth.

After scratch brushing the metal tubes are not cleaned of any debris caused by brushing. Instead, the tubes were lightly tapped to remove any loose particles. Immediately following the acetone cleaning and brushing, the tubes were stacked and extruded. On average, this was performed within 2 min to prevent the naturally occurring oxide layer to fully reform. After 15 min of exposure to air, as reported in another study [58] regarding aluminum bonding, the natural oxide layers begin to markedly interfere with the bonding process.

The scratch brushing method does not produce any noticeable debris from the stainless steel bristles which helped promote cleanliness. Other methods beyond scratch brushing were attempted but were ultimately not used. Metal files, Scotch-Brite™ pads (3M Company, Saint Paul, MA, USA), and various sanding papers were also used with no success. The major issue with these methods is cleanliness control and the lack of versatility to be applied to curved surfaces of the inner and outer diameters. These methods created a lot of non-metallic debris during application and caused too much inconvenience during processing.

2.2.2. Extrusion

The extrusion process bonds the two metals together through severe plastic deformation as described by the thin film theory. Previous research in die geometry has determined an outer ring die paired with a straight mandrel produces the least peak stresses within the die and can successfully achieve enough plastic strain to promote bonding [31]. Therefore, the die angle and geometry are adopted for this research. Figures of the die are presented in the next section. The extrusion process is performed in a 4-pronged die set which was customized to support a self-aligning die and punch. The die, punch, and mandrel are used to maintain the tubular shape and are described in more detail in the die design section. The extrusion was performed at room temperature (approximately 20.5 °C) with an extrusion speed of 2.73 mm/s. No heat was added during extrusion to promote bonding.

The die, mandrel, and punch are coated with a thin layer of extrusion oil to reduce friction and discourage material adhesion. Non-diluted Drawsol® WM 4740 (Houghton International, Manchester, England) was used for its ability to maintain high film strength when under extreme pressure. This oil is also recommended for various metals including steel, stainless-steel, titanium, and aluminized alloys. From a processing perspective, this synthetic lubricant is water soluble, which is easily removed with running tap water.

Bimetals are extruded at 52% and 68% deformation which represent how much the outer diameter is reduced during extrusion. These values represent the minimum (50%) [5–9] and mid-range of reported deformation employed in previous research, which achieved bonding using ARB. These deformation values also promote stacking as well; at 52% deformation a 2-layer tube can be re-stacked, and the original die can be reused, while at 68% deformation, 3-layer tube stacking can be utilized. This was done intentionally to reduce the number of required dies. In both cases, the deformation percentage values are slightly more than 1/2 and 2/3 as to provide clearances from the nominal stacking fraction to assist in processing.

2.2.3. Cutting

The second step in the iterative loop, if the bimetallic layer thickness is not achieved, is cutting. Simply, the bimetal is cut to remove the non-bonded section at the end of the bimetallic extrusion, the non-bonded initial section at the start, and then equally in half perpendicular to the extrusion direction. By performing cutting the total material volume is not conserved. Therefore, it is necessary to quantify the expected losses per iteration to ensure a viable end-product is produced. Cutting is performed using a material specimen preparation sawmill using a diamond infused metallurgical cutting disc. After, all edges were deburred with 320 grit sandpaper.

2.2.4. Expansion

One of the two extruded tubes require diametrical expansion to facilitate stacking. The chosen tube is expanded such that the inner diameter is increased to a size that is slightly larger than the outer diameter of the extruded tube. This is performed by pushing the tube over a diametrical expansion mandrel. The expansion mandrel utilizes a cylindrical section that tapers at 10° to an enlarged diameter. The punch, which pushed the bimetal into the extrusion die, is the same punch used to push the tube over the expansion ledge. The bimetal is passed over the expansion post multiple times to fully remove spring-back and achieve a cylindrical tube for stacking. The expansion mandrel is described in more detail in the die design section. Other options to expand the tube were considered such as metal spinning as described in [59,60]. Ultimately, it is more desirable to expand the tube using a mandrel because the process is simple, easy to control, and contained to the same experimental setup (i.e., the hydraulic press and die set).

The expansion step does not promote or impact bonding at the interface since bonding occurred during extrusion where significantly large strains (compared to strains experienced during expansion) are imposed. Additionally, from a conservation of volume perspective, the wall thickness of the expanded tube will decrease depending on the amount of deformation imposed (~11% and ~8% for the 52%, and 68% deformation cases, respectively), which will impact layer thickness consistency. Even though wall thicknesses become slightly inconsistent as iterations continue, the intent is to create many interfacing layers, i.e., the ultrafine structures independent on local layer thicknesses. The local deformation conditions also cause non-uniformity in layer thickness.

2.2.5. Heat treatment

During the extrusion and expansion processes, the metal experiences severe plastic deformation, which is causing significant strain hardening. To aid in processing, it is necessary to restore ductility. Additionally, research in ARB, which is similar to AEB, requires intermediate annealing [61]. Initial annealing of the as-received material is performed to remove initial tempers of T6 for aluminum and H58 for copper. The as-received copper is annealed to 426 °C with a 1 h soak time and cooled at a rate of 426 °C/h. The as-received aluminum is annealed to 413 °C with a 2.5 h soak time and cooled at a rate of 28 °C/h. The initial annealing was selected to enhance ductility [62].

Intermediate annealing is performed after every iteration, and the same annealing as described above for annealing aluminum was employed on the bimetal copper-aluminum tubes produced. It was found that annealing was necessary in every iteration step. Multilayered bimetals were attempted with the annealing step omitted, and severe blistering and tearing occurred throughout the tubular wall.

3. Die and Expansion Designs

The experimental setup of the extrusion process is shown in Figure 3a. The setup is installed in a hydraulic press capable of 260 kN (Greenerd Press and Machine Co., Nashua, NH, USA). The press is omitted from figures. As indicated previously, extrusion was performed at 52% and 68% radial strain. The only difference between the two extrusions is the size of the extrusion ledge (i.e. the dimension L, as shown later) to increase the strain from 52% to 68%. The off-the-shelf die set is customized to perform both the extrusion and expansion processes. The extrusion die and the expanding post are easily swapped to either perform the extrusion or expansion process. A cut-away view of the extrusion setup is provided in Figure 3b,c. Critical dimensions are shown in Figure 3c and are tabulated in Table 2.

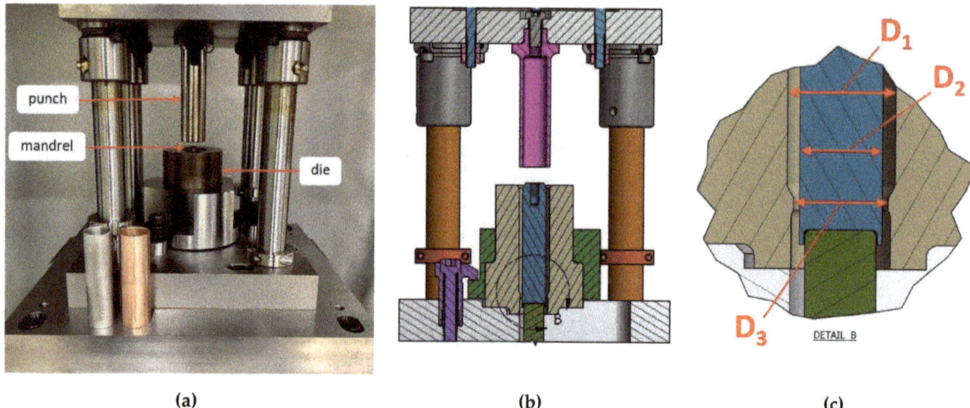

Figure 3. (**a**) Experimental setup of extrusion process which provides 52% radial strain. Omitted is the hydraulic press. Also depicted is annealed copper and aluminum test metals ready for extrusion. For size perspective, the test metals are 89 mm in height. This setup is identical to the process which enacts 68% radial strain except for dimension L. (**b**) Section view of extrusion setup. (**c**) Die cavity and extrusion ledge. The tubular bimetal is omitted.

Table 2. Critical dimensional values of extrusion die.

Deformation	D_1 mm	D_2 mm	D_3 mm
52%	28.70	22.07	25.25
68%	28.70	22.07	24.18

For either the extrusion or the expansion process, both configurations utilize the punch as the mechanism to enact deformation. This was tactically chosen to keep the setup contained to one die set installed in one hydraulic press. Both the die and the expansion post use a floating alignment method. To ensure the punch is always axially aligned neither the extrusion die, nor the expansion post are fixed in place; rather, they both self-align to the punch during setup.

The die is of sufficient length to fully encapsulate the length of the bimetal tubes. This is to ensure it is forced into the extrusion ledge. The extrusion edge geometry, which has a 30° transfer from the initial diameter to the extrusion diameter, has rounded and smooth radii. Just below the extrusion ledge is a diametrical relief for ease of tube removal after the tube is extruded into and past the extrusion ledge. The inner mandrel floats collinear to the die and remains collinear when the bimetal tubes are installed inside the die. The floating mandrel is positioned such that only the least amount of the mandrel is below the extrusion ledge to aid in the removal of the bonded bimetal tube. After extrusion is performed, due to the setup in a hydraulic press, the bimetal tube is removed by removing the die and floating mandrel. For this reason, the floating mandrel is not attached to the vertical support below it. A relief is cut into the bottom portion of the die set, the width slightly larger than an extruded tube, to assist in bimetallic tube removal. Lastly, the punch is designed to insert into the die and have the floating mandrel insert in it. Clearances between these components are less than 2.54×10^{-2} mm.

The expansion post for the expansion process, shown in Figure 4, has a smooth transitional ledge, at 10°, used to expand the bimetal tube such that it can fit over a tube of its original extruded size. A few variations of the expanding post can be used in which the expansion diameter is increased in predetermined increments to aid in processing. Incorporating these variations allows a step-up approach to achieving the final desired inner diameter of an expanded bimetal, if needed. Alternatively, the diameter may be increased directly in one expansion step. For the testing performed, all tubes were directly expanded in one step. Sitting at the base of the expanding post is an oversized washer which can be used to aid in the removal of the bimetal.

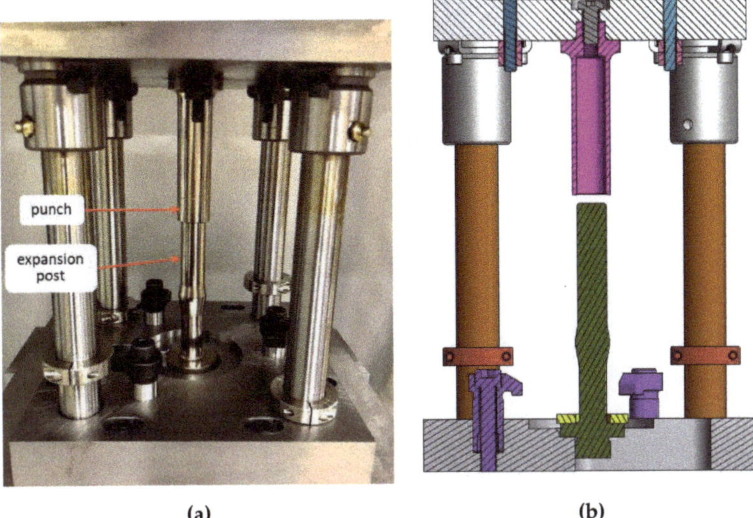

(a)   **(b)**

Figure 4. (**a**) Experimental setup of expansion process. Omitted is the hydraulic press. (**b**) Section view of expansion setup. Expansion setup uses the same die set as the extrusion process.

The die, punch, mandrel, and expansion post are the main functional components that are performing the extrusion or expansion. The material for these components is AISI A2 tool steel. This material is commonly used in extrusion dies and other high stress material forming processes. The hardness range for these components is 58 to 62 HRC which is a typical range for extrusion and forming dies. The A2 material has desirable characteristics which are tabulated in Table 3. The wear resistance and toughness are improved with the coating described below.

Table 3. Cold work tool steel relative ratings (A = greatest to E = least) [63].

Characteristic	AISI A2 Tool Steel
Safety in hardening	A
Depth of hardening	A
Resistance to decarburization	B
Stability of shape in heat treatment	A
Machinability	E
Hot hardness	C
Wear resistance	B/C
Toughness	E

The die, punch, mandrel, and expanding post are coated in a thermal diffusion process, which is a typical coating process for blanks, dies, and components used in similar high stress forming operations. The coating provides additional lubricity and reduces reactionary stresses during extrusion. Additionally, the tool toughness and hardness are promoted, and in general, the life of the components are extended. The coating data is tabulated in Table 4.

Table 4. Coating data for die, mandrel, punch, and expansion post [64].

Coating Information	Result
Thickness (μm)	5.08–7.62
Micro hardness (HV)	3500–3800
Coefficient of friction	0.08
Composition	Vanadium carbide

4. Materials

The initial aluminum tube (Ø25.40 mm × 1.65 mm) is 6061 per ASTM B210 and the initial copper tube (Ø28.58 × 1.65 mm and Ø25.40 mm × 1.65 mm) is C12200 per ASTM B75. The true stress-strain curves of the materials, after initial annealing, are presented in Figure 5. Flow curves were determined per ASTM E8 using the bulk tube as the specimen.

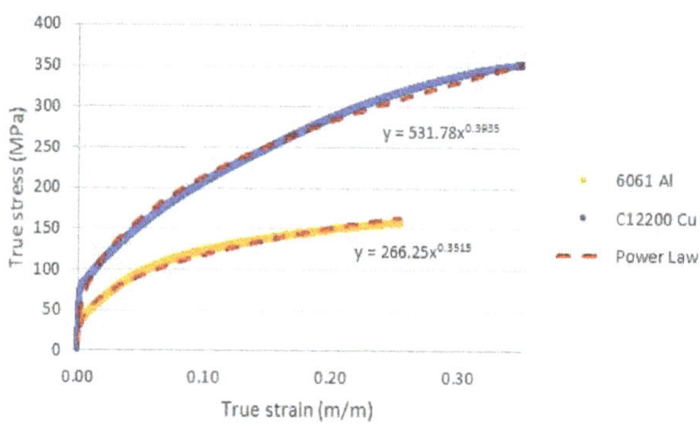

Figure 5. True stress-strain curves of annealed copper and aluminum. Material behavior is modeled by the power law.

Before processing, the as-received material is annealed since the aluminum and copper were tempered to T6, and H58, respectively. Both initial annealing cycles were performed as recommended by the Society of Manufacturing Engineers to produce optimally ductile materials as previously described. To confirm the effectiveness of the initial annealing, hardness measurements were taken; the results are shown in Table 5. The copper experienced a 60.2% decrease in hardness and the aluminum experienced a 67.8% decrease. Knoop hardness testing was performed using a 500 g force held for 10 to 15 s where the hardness value was averaged over 10 samples.

Table 5. Hardness of mill and annealed material (HK).

Material	Mill	Annealed
Copper	141.1	56.1
Aluminum	126.4	40.7

The copper and aluminum material behavior are fitted with the power law with an R^2 value of 0.9816, and 0.9893, respectively. Figure 5 shows that the copper and aluminum exhibit strain-hardening which is represented well by the power hardening law shown in Equation (1):

$$\sigma = K\varepsilon^n \tag{1}$$

The strength coefficient, K, and the strain hardening exponent, n, are presented in Table 6. An important item of note is the dissimilarity between both materials true stress-strain curves. For consistent plasticity to occur within both metals, and to maintain balanced layer formation, similar flow stress behavior across the material is tactically sought; however, the copper has a strength coefficient 99.7% larger than the aluminum, which is expected to influence the difference between extruded layer thicknesses. The values of the strain hardening exponent represent how quickly the material hardens when deforming. A value closer to zero represents a material resisting deformation, while the values closer to one represent a material where true stress and true strain vary proportionality.

Table 6. Power law strength coefficient and exponent.

Material	Strength Coefficient K, MPa	Strain Hardening Exponent n
Copper	531.78	0.3935
Aluminum	266.25	0.3515

5. Finite Element Method-Based Simulations of Extrusion

The extrusion process is modeled in ANSYS Mechanical (ANSYS 19.1., ANSYS Software Company, Canonsburg, PA, USA) to gain insight on the plastic behavior of the bimetal and to understand the stresses that develop within the die. The geometry of the extrusion model consists of four components: Two metals experiencing extrusion and two workpieces enabling the extrusion. The two metals are referred to as the outer and inner metals, which represent the initial outer, and inner diametrical layers, respectively. For all simulations and experiments, the copper is always the outermost metal tube. The model is axisymmetric and 2-dimensional. Figure 6a shows the ANSYS model of the extrusion process, where axis symmetry is taken about the farthest left edge. The geometry represents a 2-dimensional "slice" of the area-of-interest, which is the lower section of the die and mandrel identified by the red circle in Figure 6b. The model is axisymmetric and 2-dimensional because no 3-dimensional irregularities in stress or strain are expected since the tooling is precision-ground and the design utilizes a self-centering die.

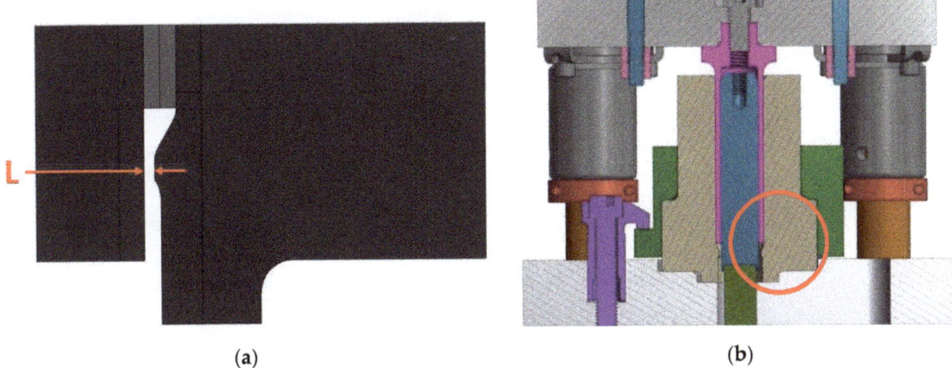

Figure 6. (**a**) Finite element geometry which represents the two deformation cases studied: 52% and 68%. Dimension L was adjusted to obtain the two different cases. (**b**) The experimental setup is shown at dead-bottom position, where the simulated area is identified by a red circle.

The geometry of the finite element model consists of the bottom section of the extrusion process. This is the section which contains the extrusion ledge at the bottom of the die. No further geometry is necessary because the model has been iteratively reduced to capture the significant stresses in the die while tolerating manageable convergence duration. The punch is not modeled; instead, an input displacement is utilized.

The model is partitioned into various sub-sections of the original geometry to focus higher element density to the area of interest which is the internal edges of the mandrel and die. The mesh is shown in Figure 7. After performing a mesh sensitivity study, the model includes 26,286 elements with 75,785 nodes.

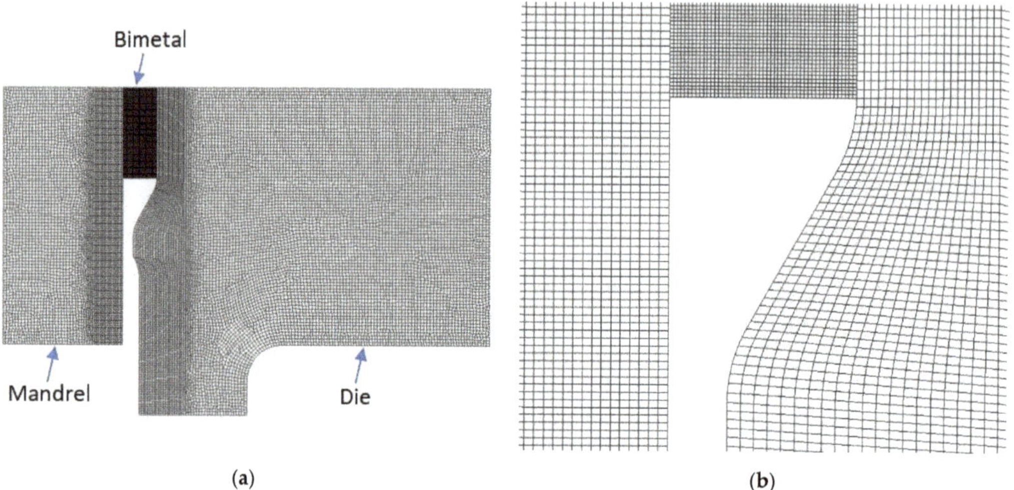

Figure 7. (**a**) The extrusion model meshing, and (**b**) close-up of the meshing, where the metals that will experience extrusion are located at the top of the die ledge at t = 0 s.

The contact control between the die, mandrel, and the metals experiencing extrusion are controlled with an augmented LaGrange formulation with nodal-normal to target detection method. The augmented LaGrange formulation comes with a computational penalty for longer solve time but controls nodal penetration very well, which is important during sliding-type simulation. An allowance of 1.27×10^{-2} mm penetration was tolerated. The contact between the two workpieces is bonded as a simplification to assist in convergence.

A frictional value of 0.025 is used between all sliding surfaces, which is consistent with the conclusions of [65], but slightly less than the values used in other research [66–68]. For comparison to rolling, this frictional value is less than the "normal" lubrication value of 11 as reported in [69]. Unlike rolling, the frictional value must be as low as possible in practice as metal adhesion is a major failure mode in extrusion and is not easily resolved as in rolling. For this reason, the A2 tool steel of the die and mandrel are polished to a 0.8 µm Ra surface finish after a thermally diffused coating of vanadium carbide is applied. The coating has a hardness of 3400 HV minimum and is very smooth. In addition, lubrication is used during testing to reduce the friction coefficient as modeled.

An input displacement of 7.62 mm is applied to the top surfaces of the metals experiencing extrusion which forces them to interact with the extrusion ledge as shown. The input of 7.62 mm is used, as this is sufficient displacement to achieve steady-state plastic flow during the extrusion simulation.

The die and mandrel are evaluated for yielding in four instances where copper-copper and copper-aluminum bimetals are extruded at 52% and 68% deformation. Results are shown in Figure 8. Peak stress occurs on the 30° ledge of the die when extruding copper-copper at 68% deformation where a maximum Von Mises stress is found to be 955 MPa. This demonstrates that a factor-of-safety of 1.7 is achieved in the most stressed condition with the yield strength of 1600 MPa estimated for the die material. Because of this, neither mandrel nor the die are expected to yield. As shown in previous research, a 30° die angle is optimal when compared to 22.5, 45, and 60° angles of similar die design for reducing peak stresses [31].

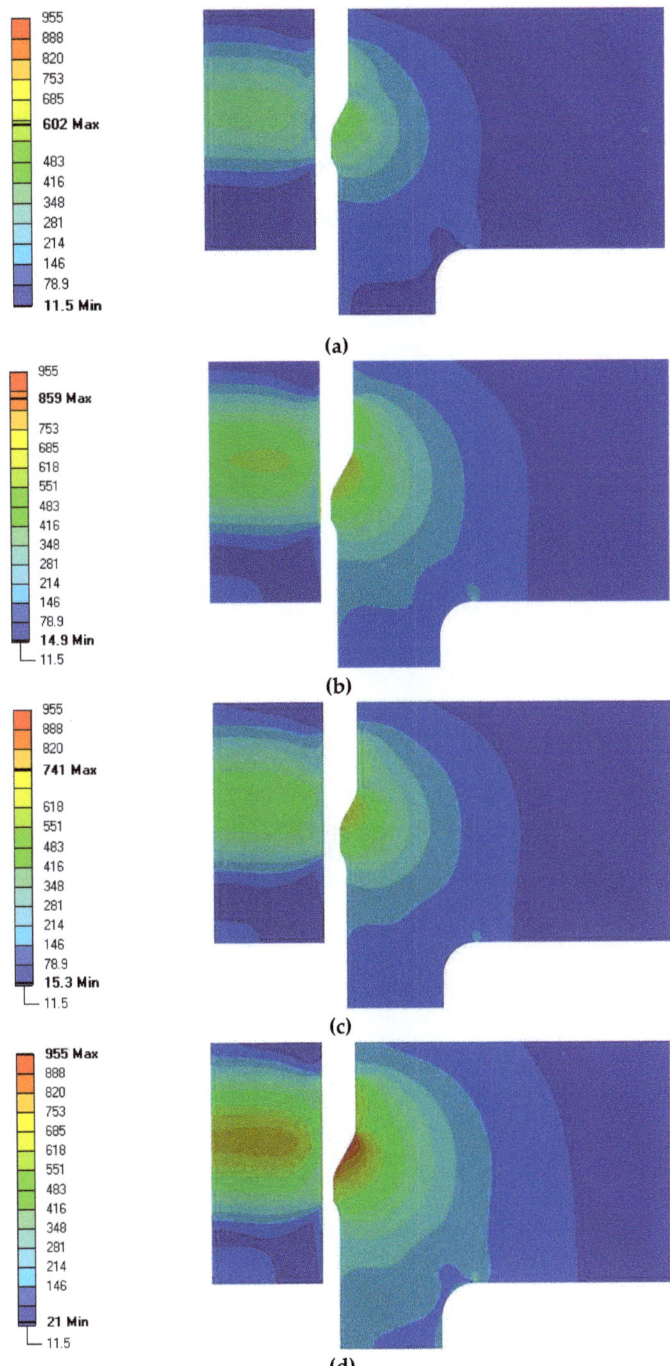

Figure 8. Von Mises stress within the die and mandrel during extrusion of copper-aluminum at (**a**) 52%, and (**b**) 68% deformation and copper-copper at (**c**) 52%, and (**d**) 68% deformation. The metals that experienced extrusion are omitted. Units of stress are MPa.

Steady-state extrusion begins after 4 mm of punch displacement. The peak input force, found at 4.4 mm, is 167 kN as shown in Figure 9 for the copper-copper simulation. Beyond this peak, the input force decreases linearly with a slight negative slope as less bimetal is within the die causing sidewall friction. As shown, the maximum input force is increased by 60.8% for the copper-aluminum bimetal when increasing the deformation from 52% to 68%, and a 55.7% input force increase is observed for the copper-copper bimetal. The predicted peak input force matches the recorded peak input forces within ~1% when averaged across multiple extrusions, where the multiple recorded peak forces are found contained within 5% of the expected peak force.

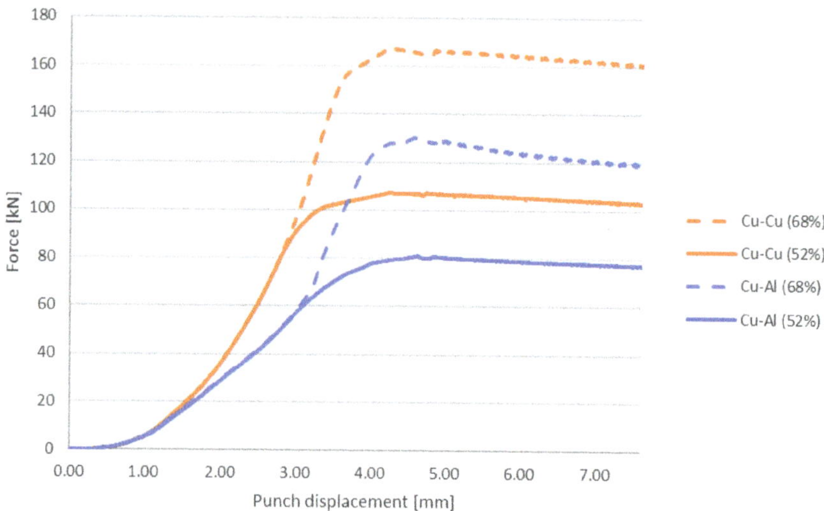

Figure 9. Force-displacement curve for extrusion at 52% and 68% deformation for copper-copper and copper-aluminum bimetals.

As the bimetals pass the extrusion ledge, a significant amount of plastic deformation occurs. The resultant total plastic deformation is shown in Figure 10 for all four cases. In the steady-state extrusion, the plastic strain varies through the thickness and length of the bimetal. Shown in Figure 11 is the plastic strain variance through the wall thickness. In all cases, a higher plastic strain is found on the innermost edge of the tube and decreases close to linearly to the midpoint. Beyond the midpoint, the plastic strain levels out but then dips close to the outer diameter. The innermost metal exhibits higher strain because the outer tube's edge is impacted by the extrusion ledge first, which forces the inner metal to push ahead of the outer metal causing more strain in the inner metal. This is observed in all four cases, and predominately in the 52% deformation case, where the inner material is drawn past the die and is extruded first (reference Figure 10a). Additionally, the layer thickness varies ± 3% from the nominal thicknesses. Most notable, as observed in Figures 10 and 11, is the drastic increase in plastic strain at the metal-metal interface: 50.1% for copper-copper and 52.6% for copper-aluminum when the deformation is increased from 52% to 68%.

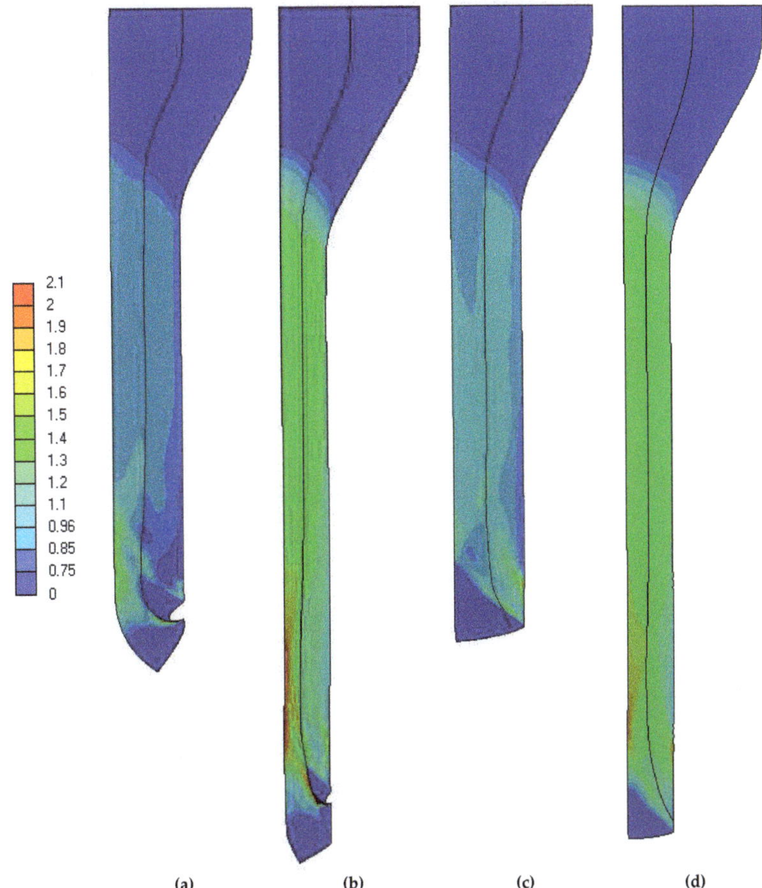

Figure 10. Total equivalent plastic strain at 7.62 mm of displacement for copper-aluminum at (**a**) 52%, and (**b**) 68% deformation and copper-copper at (**c**) 52%, and (**d**) 68% deformation. The die and mandrel are omitted. Units of strain are mm/mm.

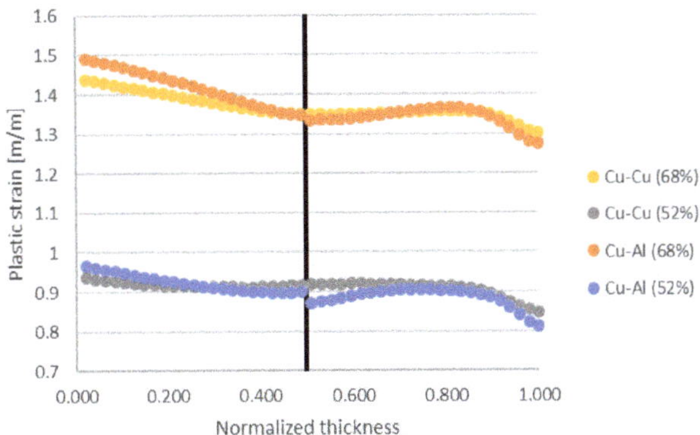

Figure 11. Equivalent plastic strain through the extruded wall where normalized 0.0 is the inner edge of the tubular wall and normalized 1.00 is the outer surface of the tube. Vertical line represents bonding interface location. Plastic strain is taken at the vertical midpoint of the extruded bimetal.

Producing the tubes in this manner requires sacrificing the beginning and end of the extruded tubes. As is evident, the initial section does not fully achieve plastic deformation and the end section is not pressed fully past the extrusion ledge.

6. Results and Discussion

Copper-copper and copper-aluminum bimetallic tubes were produced using the methodology described. The initially stacked copper-copper and copper-aluminum tubes were measured to have an average outer diameter of 28.58 mm and an inner diameter of 22.1 mm before processing. After the AEB process, the outer diameters were reduced to an average of 25.324 mm for the 52% extrusion and 24.257 mm for 68% extrusion. For all extrusions, the inner diameters were found to be 22.073 mm on average. This demonstrates that 49.9% and 66.3% radial deformation was achieved which is less than the targeted 52%, and 68% deformation, respectively. This is due to initial air gaps between the stacked layers, the mandrel, and the die, as well as expansion of the die during extrusion.

Table 7 summarizes all bimetallic tubes produced and whether bonding was achieved or not. The copper-copper tube serves as a baseline and represents the easiest possible chance to achieve full-bonding since metallographic substructures are consistent and the material exhibits identical material behavior. Copper-aluminum, however, represents a combination that is more difficult to bond due to differences in material behavior and different naturally occurring oxide layers. Four-layer tubes of copper-copper, and higher, were not attempted since the 2-layer copper-copper tube demonstrates that bonding is possible using AEB. Further processing of a copper-copper tube would provide no more desired insight as future iterations are expected to bond as the first 2-layer.

Table 7. Summary of bimetallic tubes produced.

Material	52% Deformation		
	2-layer	4-layer	8-layer
copper-aluminum	Yes *	Yes *	Yes *
copper-copper	Yes *		
Material	**68% Deformation**		
	2-layer	4-layer	8-layer
copper-copper	Yes		

* = no bonding.

Despite adhering to the methodology described above, all samples extruded at 52% did not bond. The only successful bonding occurred at 68% deformation in the copper-copper bimetallic tube. Based on the extrusions performed, deformation greater than 52% is required for bonding since the processing remained constant between all tests. The processing of the 8-layer copper-aluminum tube demonstrates that the extrusion and expansion process can create bimetallic tubes, but deformation percentage needs to be of sufficient strain to enact bonding as shown in the 68% deformed copper-copper tube. Most literature in ARB reports 50% as the low-end of deformation required to enact bonding. Evidently, bonding in the AEB process did not occur at 50% revealing the role of more complex geometry of tube relative to sheet. Mechanical fields in the tube during AEB are different than those in the sheet during ARB making the strain levels required for bonding greater in AEB process than in ARB process.

The copper-aluminum bimetal tube cross-section, shown in extrusion direction, is displayed in Figure 12. As shown in this paper, the layer thickness decreased exponentially, while the tube wall thickness remained constant. Tabulated in Table 8 are the minimum and maximum layer thicknesses measured at the cross-section taken. The 2-layer bimetal tube has very consistent layer thicknesses as well as the 4-layer bimetal. At 8-layers, the layer thickness varied greatly where some layers completely thinned to obsolesce. Beyond this, layer thickness was very inconsistent in the 8-layer bimetal.

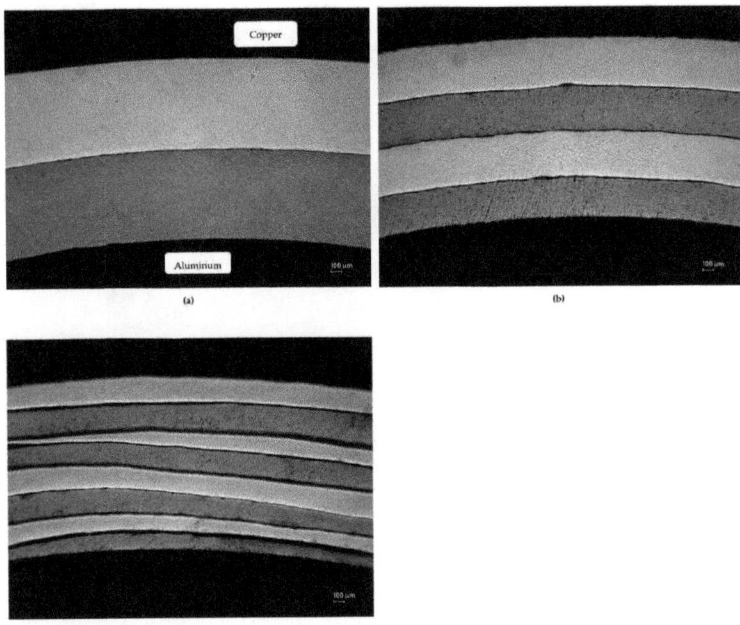

Figure 12. Bimetallic tube of (**a**) 2-layer, (**b**) 4-layer, and (**c**) 8-layer copper-aluminum, shown in the extrusion direction, produced at 52% deformation.

Table 8. Minimum and maximum layer thickness (μm) at 52% deformation.

Layers	2		4		8	
Material	**Cu**	**Al**	**Cu**	**Al**	**Cu**	**Al**
Expected	794	794	396	396	198	198
Average	819	775	414	375	212	200
Minimum	802	743	385	349	0	0
Maximum	847	798	462	402	440	504

In each sample produced at 52% deformation, bonding did not occur and is observed as the dark voids at each interface, as shown in Figure 12. Each subsequent extrusion pass did not further promote bonding as observed in the 8-layer bimetal. For this reason, it is necessary to achieve bonding on the first extrusion iteration. Since bonding did not occur, the material layers acted independently for each future extrusion, and the thin layers did not handle the imposed plastic strain, which ultimately caused significant wrinkling and tearing on the innermost and outermost layers, as well as layer thinning inside the bimetal.

Copper-copper bimetallic tubes were attempted at both 52% and 68% deformation. As mentioned previously, no bonding was achieved at 52% deformation when attempting a copper-copper bimetal. However, using the same method described, bonding was achieved using 68% deformation. As shown in Figure 13, the 2-layer copper-copper bonded interface is observed normal to the extrusion direction. Unlike Figure 12, the copper-copper cross section, shown in Figure 13, required acid-etching to view the interface using microscopy. The layer thickness was measured and found to be 510 μm for the outer layer of copper and 568 μm for the inner layer of copper where the expected thickness was 527 μm. This is attributed to the outer layer being pulled in front of the extrusion ledge during the initial extrusion start which is exhibited predominantly in Figure 10a,b.

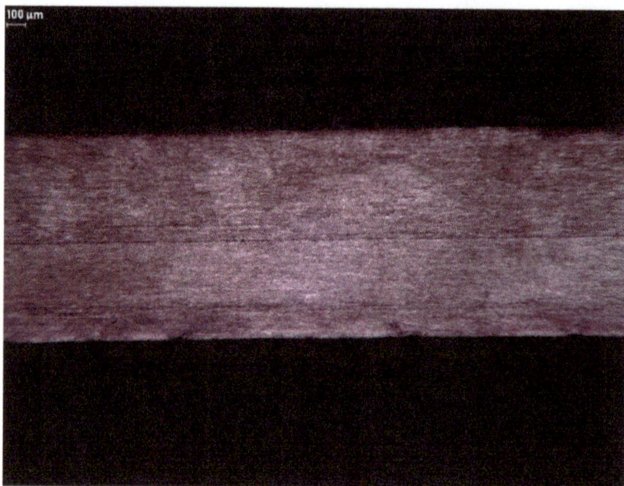

Figure 13. Bimetallic tube of 2-layer copper-copper, shown in the transverse direction, produced at 68% deformation.

Full bonding, however, did not occur as there are voids at the interface. A 15,240 µm long section was surveyed and 85.0% of the length was found to be bonded. In this section, 148 voids were identified with an average length of 15.4 µm. No identifiable pattern was observed regarding the location of the voids, and the largest void was found to be 49.7 µm. Typical voids are shown in Figure 14, which are represented by black at the interface. These voids are expected to collapse during further iterations.

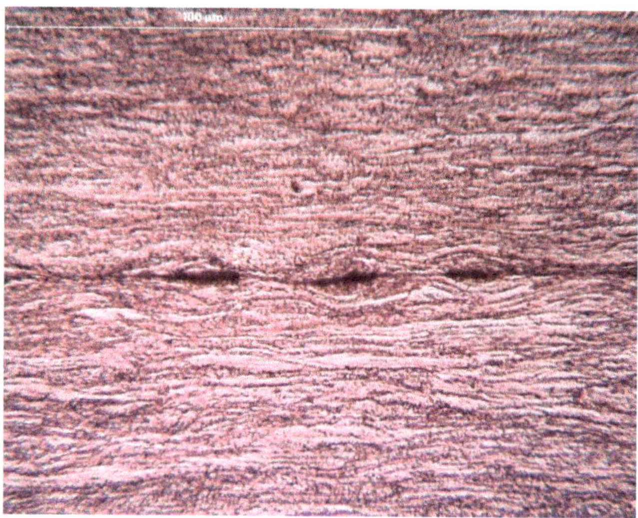

Figure 14. Typical voids found at bonding interface of 2-layer copper-copper tube shown in transverse direction. Bimetal was extruded at 68% deformation.

The grain structure before and after extrusion is displayed in Figure 15. As shown in the transverse direction, the annealed grain structure became highly elongated due to the extrusion process. The copper-copper tube after extrusion is expected to exhibit an anisotropic material behavior where grains no longer have uniformity in all spatial directions, which is expected for non-heat-treated metal after drawing or extrusion.

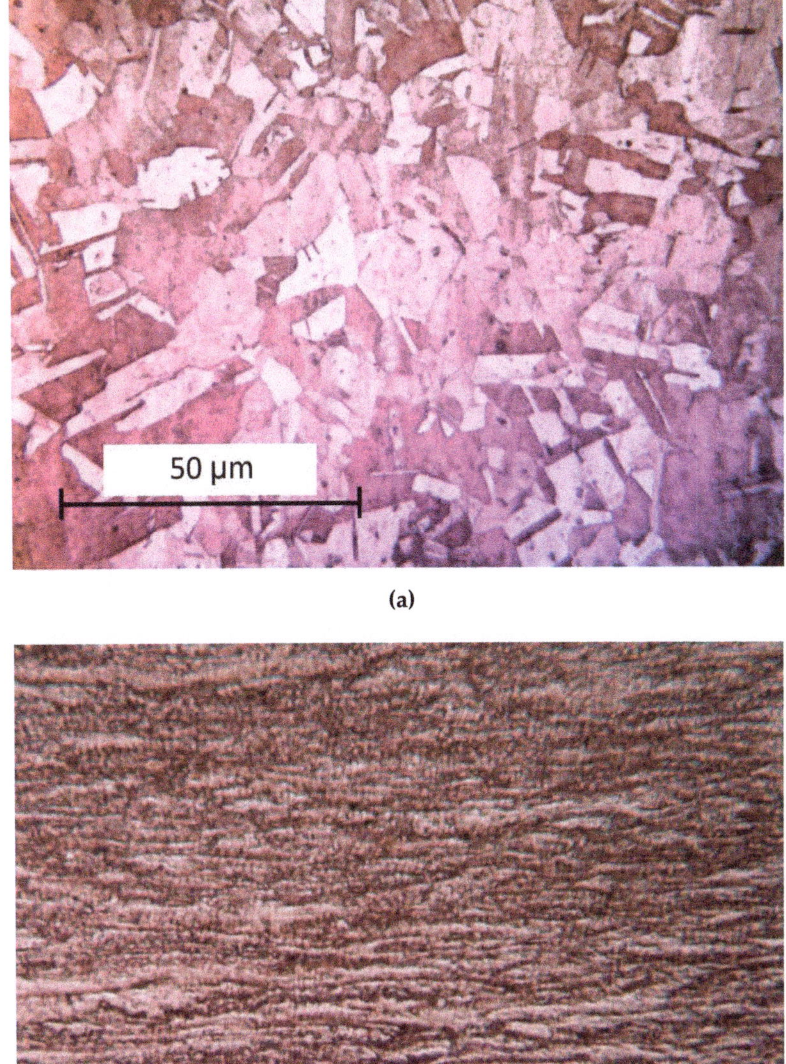

(a)

(b)

Figure 15. Microstructure of copper, (**a**) before extrusion, and (**b**) after extrusion at 68% deformation.

The copper-aluminum and copper-copper bimetal underwent significant strain-hardening during extrusion at 52% and 68% deformation. As tabulated in Tables 9 and 10, the hardness of each metal constituent increased significantly. The 8-layer copper-aluminum was not tested for hardness since the individual layers were too small for micro-hardness testing. Interestingly, the copper-copper layers experienced approximately the same increase (165.1% vs. 168.5%) in hardness even though the deformation percentage was 52% and 68% respectively. This suggests there is a hardening limit as no significant increase in hardness was observed with strain.

Table 9. Hardness (HK) before and after extrusion of 2-layer copper-copper bimetal.

Deformation	Annealed	2-Layer	Increase (%)
52%	56.1	148.8	165.1
68%	56.1	150.7	168.5

Table 10. Hardness (HK) before and after extrusion of copper-aluminum bimetal at 52% deformation.

Material	Annealed	2-Layer	Increase (%)	4-Layer	Increase (%)
Copper	56.1	144.0	156.6	146.9	161.9
Aluminum	40.7	72.8	78.9	71.8	76.4

The copper-copper tube, extruded at 68%, exhibits significantly improved material strength as indicated by the material behavior displayed in Figure 16. A second tensile test was performed to confirm results; a difference of ~2% was identified between the two ultimate tensile strengths found. Tensile tests were performed per ASTM E8 using custom sidewall specimens. The 0.2% offset yield strength improved from 83 MPa to 481 MPa; a 480% increase compared to the annealed material. Due to the work hardening experienced during extrusion, ductility is sacrificed for the improved material strength.

The ultimate tensile strength of the copper-copper tube, extruded at 68% using AEB, is compared to pure copper experiencing ARB, as reported by [70], and tube cyclic extrusion-compression (TCEC), as reported by [71]. TCEC is a severe plastic deformation technique where tubes are fully constrained and deformed between an external chamber and an internal mandrel [71]. The pure copper undergoing ARB and TCEC achieved ultrafine grain size after four passes of severe plastic deformation processes. Pure copper experiencing 68% AEB, on the other hand, has a grain size approximately one order of magnitude greater but exhibits the greatest improved ultimate tensile strength in one iteration. The significant improvement of strength is attributed to the hyper-elongated grains shown in Figure 15b, which are oriented in the extrusion direction, where dislocation structures, and underlying residual stress fields expected to form during the process similar to ARB [72]. We observe that grain size obtained is an order of magnitude larger than other severe plastic deformation processes (Table 11), but the largest improvement in ultimate tensile strength was obtained. Therefore, grain size refinement alone does not govern improvements in material strength, but also other features such as dislocation density and low angular boundaries in sub-grain structure also impact strength. Anisotropic material properties are expected since the average grain size is 2.3 µm in the extrusion direction and an average length of 40 µm is in the transverse direction.

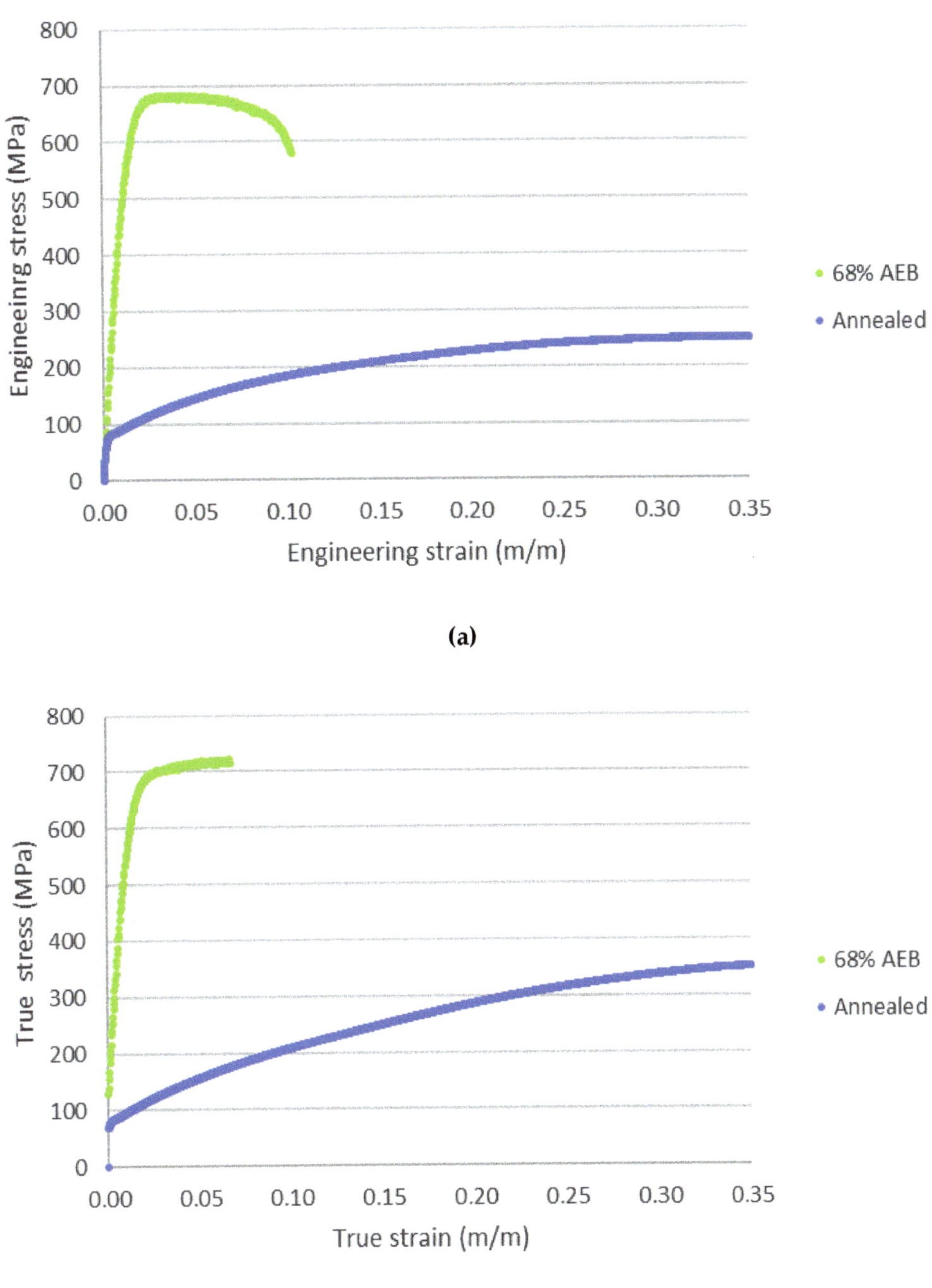

(a)

(b)

Figure 16. Engineering stress-strain, (**a**) and true stress-strain, (**b**) of 68% deformation 2-layer copper-copper compared to annealed curves.

Table 11. Ultimate tensile strength of pure copper deformed using different severe plastic deformation processes.

Process	Iteration	Ultimate Tensile Strength (MPa)	Percent Difference to Annealed (%)
68% AEB	1	683	172%
ARB [70]	1	350	39%
ARB [70]	2	370	47%
ARB [70]	3	395	57%
ARB [70]	4	395	57%
TCEC [71]	1	275	10%
TCEC [71]	2	300	20%
TCEC [71]	3	310	24%
TCEC [71]	4	325	29%
Annealed	-	251	0%

7. Summary and Conclusions

This work developed a process termed accumulative extrusion bonding for introducing laminated structures in metallic tubes. To this end, dies are designed and evaluated using the finite element method simulations. Several multilayered tubes are produced using the process and characterized for the extent of bonding, microstructure, hardness, and strength. Significantly, bonding at the interface is achieved for the copper-copper metallic tubing to about 85% at a radial strain of 68% imparted by the AEB process. Since complete bonding is desired, it is recommended to increase the radial deformation to a value greater than 68%. The additional key findings are:

1. Bonding using AEB does not occur at 50% deformation revealing the significant role of more complex geometry of tubes relative to sheets in solid-state bonding. Mechanical fields in the tube during AEB are different from those in the sheet during ARB making the required strain levels for bonding greater in AEB process than those in ARB process.
2. It is necessary to achieve bonding on the first extrusion pass/iteration as future extrusion passes would not promote bonding. Since bonding does not occur on the first extrusion pass at 52% deformation, the material layers act independently for each future extrusion. Moreover, the layers begin to lose their integrity with plastic strain.
3. Surface preparation before forming the interface is very important to facilitate bonding using AEB. Any imperfection left on the interfacial layer will become an inclusion at the interface. During processing it is therefore critical to minimize inadvertent mishandling or extraneous debris. Further processing will thin and stretch areas of contamination but will not remove the inclusions. Moreover, oxidation layer minimization is necessary to aid in bonding. Scratch brushing is used as the process to promote surface hardening while also aiding in the removal of any oxide layer. This method also does not produce any noticeable debris from the bristles which helps promote cleanliness. It is found that over-brushing does not improve the amount of bonding. Beyond scratch brushing, which is a key application to remove oxides and encourage surface hardness, minimizing contact with the atmosphere is also essential.
4. Annealing during each iteration is necessary to remove strain hardening caused during extrusion. Samples extruded with the annealing step omitted failed during extrusion due to wrinkling and tearing.

To achieve unique material properties permitted by ultrafine-laminated structures in tubes like those achieved in sheets, it is necessary to bond different material combinations and further push the layering to the thickness at and below the grain size-level. Future work will explore the possibility to extrusion bond more metal-metal combinations, push the processing to achieve finer layering, and, as necessary, improve the die design.

Author Contributions: Conceptualization, M.R.S. and M.K.; methodology, M.R.S. and M.K.; writing—original draft preparation, M.R.S.; writing—review and editing, M.K. Both authors have read and agreed to the published version of the manuscript.

Funding: This research was funded by the U.S. National Science Foundation (NSF) under Grant No. CMMI-1727495.

Data Availability Statement: The data presented in this study are available on request from the corresponding author. The data cannot be shared at this time due to technical or time limitations.

Acknowledgments: University of New Hampshire for laboratory support and L&S Machine Company LLC for manufacturing die components. The authors also acknowledge support and input from professors Brad L. Kinsey, Jinjin Ha, and Yannis P. Korkolis.

Conflicts of Interest: The authors declare no conflict of interest.

References

1. Bimetallic Tubes. Available online: http://www.beca-engineering.com/product/bimetallictubes.php (accessed on 12 January 2021).
2. Bimetallic Tubes. Available online: https://www.dmitubes.com/bimetallic-tubes.html (accessed on 12 January 2021).
3. New Generation Bimetallic Tubes. Available online: https://www.energetictubes.com/ (accessed on 12 January 2021).
4. Bimetallic Tubes. Available online: https://www.materials.sandvik/en-us/products/tube-pipe-fittings-and-flanges/tubular-products/bimetallic-tubes/ (accessed on 12 January 2021).
5. Hosseini, M.; Pardis, N.; Manesh, H.D.; Abbasi, M.; Kim, D.-I. Structural characteristics of cu/ti bimetal composite produced by accumulative roll-bonding (arb). *Mater. Des.* **2017**, *113*, 128–136. [CrossRef]
6. Jamaati, R.; Toroghinejad, M.R. Manufacturing of high-strength aluminum/alumina composite by accumulative roll bonding. *Mater. Sci. Eng. A* **2010**, *527*, 4146–4151. [CrossRef]
7. Dehsorkhi, R.N.; Qods, F.; Tajally, M. Investigation on microstructure and mechanical properties of al–zn composite during accumulative roll bonding (arb) process. *Mater. Sci. Eng. A* **2011**, *530*, 63–72. [CrossRef]
8. Chang, H.; Zheng, M.; Xu, C.; Fan, G.; Brokmeier, H.; Wu, K. Microstructure and mechanical properties of the mg/al multilayer fabricated by accumulative roll bonding (arb) at ambient temperature. *Mater. Sci. Eng. A* **2012**, *543*, 249–256. [CrossRef]
9. Mahdavian, M.; Ghalandari, L.; Reihanian, M. Accumulative roll bonding of multilayered cu/zn/al: An evaluation of microstructure and mechanical properties. *Mater. Sci. Eng. A* **2013**, *579*, 99–107. [CrossRef]
10. Ghalandari, L.; Mahdavian, M.; Reihanian, M. Microstructure evolution and mechanical properties of cu/zn multilayer processed by accumulative roll bonding (arb). *Mater. Sci. Eng. A* **2014**, *593*, 145–152. [CrossRef]
11. Knezevic, M.; Nizolek, T.; Ardeljan, M.; Beyerlein, I.J.; Mara, N.A.; Pollock, T.M. Texture evolution in two-phase zr/nb lamellar composites during accumulative roll bonding. *Int. J. Plast.* **2014**, *57*, 16–28. [CrossRef]
12. Ardeljan, M.; Beyerlein, I.J.; Knezevic, M. A dislocation density based crystal plasticity finite element model: Application to a two-phase polycrystalline hcp/bcc composites. *J. Mech. Phys. Solids* **2014**, *66*, 16–31. [CrossRef]
13. Savage, D.J.; Beyerlein, I.J.; Mara, N.A.; Vogel, S.C.; McCabe, R.J.; Knezevic, M. Microstructure and texture evolution in mg/nb layered materials made by accumulative roll bonding. *Int. J. Plast.* **2020**, *125*, 1–26. [CrossRef]
14. Leu, B.; Savage, D.J.; Wang, J.; Alam, M.E.; Mara, N.A.; Kumar, M.A.; Carpenter, J.S.; Vogel, S.C.; Knezevic, M.; Decker, R.; et al. Processing of dilute mg–zn–mn–ca alloy/nb multilayers by accumulative roll bonding. *Adv. Eng. Mater.* **2020**, *22*, 1900673. [CrossRef]
15. Mashhadi, A.; Atrian, A.; Ghalandari, L. Mechanical and microstructural investigation of zn/sn multilayered composites fabricated by accumulative roll bonding (arb) process. *J. Alloys Compd.* **2017**, *727*, 1314–1323. [CrossRef]
16. Clemens, B.; Kung, H.; Barnett, S. Structure and strength of multilayers. *MRS Bull.* **1999**, *24*, 20–26. [CrossRef]
17. Mara, N.; Bhattacharyya, D.; Dickerson, P.; Hoagland, R.; Misra, A. Deformability of ultrahigh strength 5 nm cu/nb nanolayered composites. *Appl. Phys. Lett.* **2008**, *92*, 231901. [CrossRef]
18. Zhang, X.; Misra, A.; Wang, H.; Shen, T.; Nastasi, M.; Mitchell, T.; Hirth, J.; Hoagland, R.; Embury, J. Enhanced hardening in cu/330 stainless steel multilayers by nanoscale twinning. *Acta Mater.* **2004**, *52*, 995–1002. [CrossRef]
19. Ardeljan, M.; Knezevic, M.; Jain, M.; Pathak, S.; Kumar, A.; Li, N.; Mara, N.A.; Baldwin, J.K.; Beyerlein, I.J. Room temperature deformation mechanisms of mg/nb nanolayered composites. *J. Mater. Res.* **2018**, *33*, 1311–1332. [CrossRef]
20. Ardeljan, M.; Savage, D.J.; Kumar, A.; Beyerlein, I.J.; Knezevic, M. The plasticity of highly oriented nano-layered zr/nb composites. *Acta. Mater.* **2016**, *115*, 189–203. [CrossRef]
21. Su, Y.; Ardeljan, M.; Knezevic, M.; Jain, M.; Pathak, S.; Beyerlein, I.J. Elastic constants of pure body-centered cubic mg in nanolaminates. *Comput. Mater. Sci.* **2020**, *174*, 109501. [CrossRef]
22. Knezevic, M.; Zecevic, M.; Beyerlein, I.J.; Lebensohn, R.A. A numerical procedure enabling accurate descriptions of strain rate-sensitive flow of polycrystals within crystal visco-plasticity theory. *Comput. Methods Appl. Mech. Eng.* **2016**, *308*, 468–482. [CrossRef]

23. Knezevic, M.; Kalidindi, S.R. Fast computation of first-order elastic-plastic closures for polycrystalline cubic-orthorhombic microstructures. *Comput. Mater. Sci.* **2007**, *39*, 643–648. [CrossRef]
24. Jahedi, M.; Ardjmand, E.; Knezevic, M. Microstructure metrics for quantitative assessment of particle size and dispersion: Application to metal-matrix composites. *Powder Technol.* **2017**, *311*, 226–238. [CrossRef]
25. Misra, A.; Hoagland, R. Effects of elevated temperature annealing on the structure and hardness of copper/niobium nanolayered films. *J. Mater. Res.* **2005**, *20*, 2046–2054. [CrossRef]
26. Bellou, A.; Scudiero, L.; Bahr, D. Thermal stability and strength of mo/pt multilayered films. *J. Mater. Sci.* **2010**, *45*, 354–362. [CrossRef]
27. Han, W.; Misra, A.; Mara, N.; Germann, T.; Baldwin, J.; Shimada, T.; Luo, S. Role of interfaces in shock-induced plasticity in cu/nb nanolaminates. *Philos. Mag.* **2011**, *91*, 4172–4185. [CrossRef]
28. Wei, Q.; Li, N.; Mara, N.; Nastasi, M.; Misra, A. Suppression of irradiation hardening in nanoscale v/ag multilayers. *Acta Mater.* **2011**, *59*, 6331–6340. [CrossRef]
29. Zhernenkov, M.; Jablin, M.S.; Misra, A.; Nastasi, M.; Wang, Y.; Demkowicz, M.J.; Baldwin, J.K.; Majewski, J. Trapping of implanted he at cu/nb interfaces measured by neutron reflectometry. *Appl. Phys. Lett.* **2011**, *98*, 241913. [CrossRef]
30. Lesuer, D.; Syn, C.; Sherby, O.; Wadsworth, J.; Lewandowski, J.; Hunt, W. Mechanical behaviour of laminated metal composites. *Int. Mater. Rev.* **1996**, *41*, 169–197. [CrossRef]
31. Knezevic, M.; Jahedi, M.; Korkolis, Y.P.; Beyerlein, I.J. Material-based design of the extrusion of bimetallic tubes. *Comput. Mater. Sci.* **2014**, *95*, 63–73. [CrossRef]
32. Valiev, R.Z.; Estrin, Y.; Horita, Z.; Langdon, T.G.; Zechetbauer, M.J.; Zhu, Y.T. Producing bulk ultrafine-grained materials by severe plastic deformation. *JOM* **2006**, *58*, 33–39. [CrossRef]
33. Knezevic, M.; Bhattacharyya, A. Characterization of microstructure in nb rods processed by rolling: Effect of grooved rolling die geometry on texture uniformity. *Int. J. Refract. Met. Hard Mater.* **2017**, *66*, 44–51. [CrossRef]
34. Kalidindi, S.R.; Duvvuru, H.K.; Knezevic, M. Spectral calibration of crystal plasticity models. *Acta Mater.* **2006**, *54*, 1795–1804. [CrossRef]
35. Jahedi, M.; Knezevic, M.; Paydar, M. High-pressure double torsion as a severe plastic deformation process: Experimental procedure and finite element modeling. *J. Mater. Eng. Perform.* **2015**, *24*, 1471–1482. [CrossRef]
36. Jahedi, M.; Beyerlein, I.J.; Paydar, M.H.; Knezevic, M. Effect of grain shape on texture formation during severe plastic deformation of pure copper. *Adv. Eng. Mater.* **2018**, *20*, 1600829.
37. Jahedi, M.; Beyerlein, I.J.; Paydar, M.H.; Zheng, S.; Xiong, T.; Knezevic, M. Effects of pressure and number of turns on microstructural homogeneity developed in high-pressure double torsion. *Metall. Mater. Trans. A* **2017**, *48*, 1249–1263. [CrossRef]
38. Jahedi, M.; Paydar, M.H.; Knezevic, M. Enhanced microstructural homogeneity in metal-matrix composites developed under high-pressure-double-torsion. *Mater. Charact.* **2015**, *104*, 92–100. [CrossRef]
39. Jahedi, M.; Paydar, M.H.; Zheng, S.; Beyerlein, I.J.; Knezevic, M. Texture evolution and enhanced grain refinement under high-pressure-double-torsion. *Mater. Sci. Eng. A* **2014**, *611*, 29–36. [CrossRef]
40. Qiao, X.G.; Gao, N.; Starink, M.J. A model of grain refinement and strengthening of al alloys due to cold severe plastic deformation. *Philos. Mag.* **2012**, *92*, 446–470. [CrossRef]
41. Bachmaier, A.; Pippan, R. Generation of metallic nanocomposites by severe plastic deformation. *Int. Mater. Rev.* **2013**, *58*, 41–62. [CrossRef]
42. Xu, C.; Xia, K.; Langdon, T.G. The role of back pressure in the processing of pure aluminum by equal-channel angular pressing. *Acta Mater.* **2007**, *55*, 2351–2360. [CrossRef]
43. Harsha, R.; Kulkarni, V.M.; Babu, B.S. Severe plastic deformation-a review. *Mater. Today Proc.* **2018**, *5*, 22340–22349. [CrossRef]
44. Valiev, R.; Korznikov, A.; Mulyukov, R. Structure and properties of ultrafine-grained materials produced by severe plastic deformation. *Mater. Sci. Eng. A* **1993**, *168*, 141–148. [CrossRef]
45. Sadasivan, N.; Balasubramanian, M. Severe plastic deformation of tubular materials–process methodology and its influence on mechanical properties–a review. *Mater. Today Proc.* **2021**. [CrossRef]
46. Zare, H.; Jahedi, M.; Toroghinejad, M.R.; Meratian, M.; Knezevic, M. Microstructure and mechanical properties of carbon nanotubes reinforced aluminum matrix composites synthesized via equal-channel angular pressing. *Mater. Sci. Eng. A* **2016**, *670*, 205–216. [CrossRef]
47. Zare, H.; Jahedi, M.; Toroghinejad, M.R.; Meratian, M.; Knezevic, M. Compressive, shear, and fracture behavior of cnt reinforced al matrix composites manufactured by severe plastic deformation. *Mater. Des.* **2016**, *106*, 112–119. [CrossRef]
48. Chen, X.; Huang, G.; Liu, S.; Han, T.; Jiang, B.; Tang, A.; PAN, F.; ZHU, Y. Grain refinement and mechanical properties of pure aluminum processed by accumulative extrusion bonding. *Trans. Nonferrous Met. Soc. China* **2019**, *29*, 437–447. [CrossRef]
49. Chen, X.; Xia, D.; Zhang, J.; Huang, G.; Liu, K.; Tang, A.; Jiang, B.; Pan, F. Ultrafine-grained al–zn–mg–cu alloy processed via cross accumulative extrusion bonding and subsequent aging: Microstructure and mechanical properties. *J. Alloys Compd.* **2020**, *846*, 156306. [CrossRef]
50. Chen, X.; Zhang, J.; Xia, D.; Huang, G.; Liu, K.; Jiang, B.; Tang, A.; Pan, F. Microstructure and mechanical properties of 1060/7050 laminated composite produced via cross accumulative extrusion bonding and subsequent aging. *J. Alloys Compd.* **2020**, *826*, 154094. [CrossRef]

51. Liu, S.; Zhang, J.; Chen, X.; Huang, G.; Xia, D.; Tang, A.; Zhu, Y.; Jiang, B.; Pan, F. Improving mechanical properties of heterogeneous mg-gd alloy laminate via accumulated extrusion bonding. *Mater. Sci. Eng. A* **2020**, *785*, 139324. [CrossRef]

52. Mehr, V.Y.; Toroghinejad, M.R.; Rezaeian, A. The effects of oxide film and annealing treatment on the bond strength of al–cu strips in cold roll bonding process. *Mater. Des.* **2014**, *53*, 174–181. [CrossRef]

53. Ana, S.A.; Reihanian, M.; Lotfi, B. Accumulative roll bonding (arb) of the composite coated strips to fabricate multi-component al-based metal matrix composites. *Mater. Sci. Eng. A* **2015**, *647*, 303–312. [CrossRef]

54. Mozaffari, A.; Hosseini, M.; Manesh, H.D. Al/ni metal intermetallic composite produced by accumulative roll bonding and reaction annealing. *J. Alloys Compd.* **2011**, *509*, 9938–9945. [CrossRef]

55. Jamaati, R.; Toroghinejad, M.R. Effect of friction, annealing conditions and hardness on the bond strength of al/al strips produced by cold roll bonding process. *Mater. Des.* **2010**, *31*, 4508–4513. [CrossRef]

56. Chitkara, N.; Aleem, A. Extrusion of axi-symmetric bi-metallic tubes from solid circular billets: Application of a generalised upper bound analysis and some experiments. *Int. J. Mech. Sci.* **2001**, *43*, 2833–2856. [CrossRef]

57. Chaudhari, G.; Andhale, S.; Patil, N. Experimental evaluation of effect of die angle on hardness and surface finish of cold forward extrusion of aluminum. *Int. J. Emerg. Technol. Adv. Eng.* **2012**, *2*, 334–338.

58. Vaidyanath, L.; Nicholas, M.; Milner, D. Pressure welding by rolling. *Br. Weld. J.* **1959**, *6*, 13–28.

59. Hua, F.; Yang, Y.; Zhang, Y.; Guo, M.; Guo, D.; Tong, W.; Hu, Z. Three-dimensional finite element analysis of tube spinning. *J. Mater. Process. Technol.* **2005**, *168*, 68–74. [CrossRef]

60. Xia, Q.; Xie, S.W.; Huo, Y.; Ruan, F. Numerical simulation and experimental research on the multi-pass neck-spinning of non-axisymmetric offset tube. *J. Mater. Process. Technol.* **2008**, *206*, 500–508. [CrossRef]

61. Ardeljan, M.; Knezevic, M.; Nizolek, T.; Beyerlein, I.J.; Mara, N.A.; Pollock, T.M. A study of microstructure-driven strain localizations in two-phase polycrystalline hcp/bcc composites using a multi-scale model. *Int. J. Plast.* **2015**, *74*, 35–57. [CrossRef]

62. Dallas, D.B. *Tool and Manufacturing Engineers Handbook*, 3rd ed.; McGraw Hill Inc.: New York, NY, USA, 1976.

63. McCauley, E.O.A.C.J. *Tool Steels*, 28th ed.; Industrial Press Inc.: New York, NY, USA, 2008.

64. Thermal Diffusion (VCTDH) Coatings. Available online: https://www.ibccoatings.com/thermal-diffusion-td-coatings (accessed on 12 January 2021).

65. de Moraes Costa, A.L.; da Silva, U.S.; Valberg, H.S. On the friction conditions in fem simulations of cold extrusion. *Procedia Manuf.* **2020**, *47*, 231–236. [CrossRef]

66. Zecevic, M.; McCabe, R.J.; Knezevic, M. A new implementation of the spectral crystal plasticity framework in implicit finite elements. *Mech. Mater.* **2015**, *84*, 114–126. [CrossRef]

67. Knezevic, M.; Beyerlein, I.J. Multiscale modeling of microstructure-property relationships of polycrystalline metals during thermo-mechanical deformation. *Adv. Eng. Mater.* **2018**, *20*, 1700956. [CrossRef]

68. Knezevic, M.; Kalidindi, S.R. Crystal plasticity modeling of microstructure evolution and mechanical fields during processing of metals using spectral databases. *JOM* **2017**, *69*, 830–838. [CrossRef]

69. Madaah-Hosseini, H.; Kokabi, A. Cold roll bonding of 5754-aluminum strips. *Mater. Sci. Eng. A* **2002**, *335*, 186–190. [CrossRef]

70. Fattah-Alhosseini, A.; Imantalab, O.; Mazaheri, Y.; Keshavarz, M. Microstructural evolution, mechanical properties, and strain hardening behavior of ultrafine grained commercial pure copper during the accumulative roll bonding process. *Mater. Sci. Eng. A* **2016**, *650*, 8–14. [CrossRef]

71. Babaei, A.; Mashhadi, M. Tubular pure copper grain refining by tube cyclic extrusion–compression (tcec) as a severe plastic deformation technique. *Prog. Nat. Sci. Mater. Int.* **2014**, *24*, 623–630. [CrossRef]

72. Alvandi, H.; Farmanesh, K. Microstructural and mechanical properties of nano/ultra-fine structured 7075 aluminum alloy by accumulative roll-bonding process. *Procedia Mater. Sci.* **2015**, *11*, 17–23. [CrossRef]

Article

Deformation Behavior Causing Excessive Thinning of Outer Diameter of Micro Metal Tubes in Hollow Sinking

Takuma Kishimoto [1,2,*], Hayate Sakaguchi [1], Saki Suematsu [1], Kenichi Tashima [3], Satoshi Kajino [4], Shiori Gondo [4] and Shinsuke Suzuki [2,5,6]

1   Department of Applied Mechanics, Graduate School of Fundamental Science and Engineering, Waseda University, 3-4-1 Okubo, Shinjuku, Tokyo 169-8555, Japan; jg-aghsk@akane.waseda.jp (H.S.); suematsu_s@asagi.waseda.jp (S.S.)

2   Kagami Memorial Research Institute of Materials Science and Technology, Waseda University, 2-8-26 Nishi-waseda, Shinjuku, Tokyo 169-0051, Japan; suzuki-s@waseda.jp

3   Factory-Automation Electronics Inc., 1-6-14 Higashi-nakajima, Higashi-yodogawa, Osaka 533-0033, Japan; k-tashima@fae.jp

4   Advanced Manufacturing Research Institute Department of Electronics and Manufacturing, National Institute of Advanced Industrial Science and Technology (AIST), Tsukuba East, 1-2-1 Namiki, Tsukuba, Ibaraki 305-8564, Japan; kajino-satoshi@aist.go.jp (S.K.); shiori-gondo@aist.go.jp (S.G.)

5   Department of Materials and Science, Faculty of Science and Engineering, Waseda University, 3-4-1 Okubo, Shinjuku, Tokyo 169-8555, Japan

6   Department of Applied Mechanics and Aerospace Engineering, Faculty of Science and Engineering, Waseda University, 3-4-1 Okubo, Shinjuku, Tokyo 169-8555, Japan

*   Correspondence: kishimoto-5389@asagi.waseda.jp; Tel.: +81-3-5286-8126

Received: 25 August 2020; Accepted: 27 September 2020; Published: 1 October 2020

Abstract: The deformation behavior of microtubes during hollow sinking was investigated to clarify the mechanism of the excessive thinning of their outer diameters. Stainless-steel, copper, and aluminum alloy tubes were drawn without an inner tool to evaluate the effect of Lankford values on outer diameter reduction. Drawing stress and stress-strain curves were obtained to evaluate the yielding behavior during hollow sinking. The observed yielding behavior indicated that the final outer diameter of the drawn tube was always smaller than the die diameter due to the uniaxial tensile deformation starting from the die approach end even though the drawing stress was in the elastic range. The results of a loading-unloading tensile test demonstrated that the strain remained even after unloading. Therefore, the outer diameter is considered to become smaller than the die diameter during hollow sinking due to microscopic yielding at any Lankford value. Furthermore, the outer diameter becomes smaller than the die diameter as the Lankford value increases, as theorized. As the drawing stress decreases or the apparent elastic modulus of the stress-strain curve increases, the outer diameter seems to approach the die diameter during unloading, which is caused by the elastic recovery outside the microscopic yielding region.

Keywords: micro tube; hollow sinking; plastic anisotropy; surface quality; size effect

---

1. Introduction

Micro metal tubes are used in various fields that entail medical devices and heat exchangers [1,2]. Thin tube walls are required to increase the volume flow rate of a liquid from needles during injection [3]. Low surface roughness of the tubes is also required to improve blood flow in injection needles when injecting liquid. Therefore, control over the wall thickness and outer surface quality are requirements of the micro tube fabrication process. A plug or mandrel is generally used to control the wall thickness during tube drawing. However, inserting a plug or mandrel into micro metal tubes is impractical for creating micro metal tubes with a maximum outer diameter of 3 mm [4]. In conventional hollow sinking, where only the drawing speed on the die's exit side can be controlled, the wall thickness generally increases. Therefore, fabricating thin-walled tubes using conventional hollow sinking is difficult [5]. Tube volume entering the die is equal to that exiting the die over a single unit of time. Therefore, the final wall thickness decreases as the length of the drawn tubes increases. In other words, the final wall thickness decreases as the drawing speed ratio on the die's entrance and exit sides increases, when the outer diameter of the drawn tube matches the die diameter during the hollow sinking. This theory for controlling wall thickness has not been validated because only the drawing speed on the die's exit side can be controlled in a conventional drawing machine. However, a contemporary drawing machine that controls both the drawing speeds on the die's entrance and exit sides has recently been developed [6]. Previously, the authors investigated the conditions for wall thickness reduction by using this machine [7]. The results demonstrated that the wall thickness could be decreased when the drawing speed ratio was larger than a threshold value, which was obtained from the die reduction and the starting dimensions. Therefore, the drawing speed ratio must be set above this threshold to reduce the wall thickness during hollow sinking.

In our previous work, a theory for controlling wall thickness during hollow sinking was established [7]. Furthermore, a high dimension accuracy is required so that the outer diameter of the drawn micro tube matches the die diameter to improve the expansibility of stents [8]. In this study, tubes with a maximum outer diameter of 2 mm, which are required for several applications [5], were defined as the microscale tubes. Generally, the outer diameter of a macroscale drawn tube, such as an outer diameter of about 20 mm, matches the die diameter [9]. However, the authors reported that the outer diameter of the microscale drawn tubes with the outer diameter of about 1.5 mm became smaller than the die diameter starting from the die approach end [10]. Furthermore, the free surface roughening developed on the outer surface of the drawn tube due to the excessive thinning of the outer diameter. We assume that this excessive thinning of the outer diameter is caused by a uniaxial tensile deformation starting from the die approach end. In this study, we focused on the following aspects to clarify the mechanism causing the excessive thinning of the outer diameter: (1) flow stress, (2) friction force, and (3) plastic anisotropy. The details are described in the following paragraphs.

Several studies have reported that the size effect caused by miniaturization influences the deformation behavior of micro tubes [11,12]. For example, flow stress decreases with the grain number across the thickness of a specimen due to miniaturization [13]. The decrease in flow stress can be explained by using a surface layer model [14]. Dislocations pile up not at the free surface grain, but at the grain boundary. Therefore, dislocation movement in the free surface grains is less obstructed than movement at core grains. Furthermore, free surface grains exhibit lesser hardening than inner volume grains [15]. The fraction of grains representing the surface layer increases with miniaturization. Therefore, the flow stress decreases with the number of grains across the thickness. Several papers reported that this size effect appeared when only 10–20 grains exist in the thickness [11,12]. According to the above discussion, we assume that a microscale tube seems to yield more easily than a macroscale tube for a given drawing stress when the tube is stretched from the die approach end. The magnitude of the drawing stress during hollow sinking can be investigated using a finite element method (FEM) [16]. However, investigating complex phenomena such as the size effect by using this method is difficult. Therefore, the yielding behavior due to the size effect should be investigated by evaluating the magnitude of the measured drawing stress against the bulk yield stress.

The size effect on friction should also be considered when verifying the mechanism of the excessive thinning of the outer diameter during microscale hollow sinking. The force required to deform the tube decreases because of miniaturization even though the frictional force does not change. The proportion of the friction force to the drawing force increases with miniaturization. As a result, the micro tube seems to yield more easily during drawing than macroscale tubes because the drawing stress applied to the micro tube increases. The outer diameter of the micro tube seems to become smaller than the die diameter because of both the size effect on the flow stress and the effect of friction.

In addition to considering the size effect, investigating plastic anisotropy is expected to help clarify the mechanism causing the excessive thinning of the outer diameter during microscale hollow sinking. Several studies have reported on the effect of plastic anisotropy on the mechanical property of sheet metal forming by evaluating Lankford values [17]. The outer diameter of tubes decreases as the Lankford value increases in a tensile test. Therefore, the outer diameter is estimated to decrease from the die approach end as the Lankford value increases during hollow sinking. Generally, this value is calculated as the strain ratio of width to sheet thickness obtained via tensile testing. A study reported a method of measuring the Lankford value of tubes by realizing tensile testing on a specimen retaining a tube-shape [18]. The Lankford value of the tube can be calculated as the ratio of the circumferential strain ε_θ to wall thickness strain ε_t. From the above information, a more extreme reduction in the outer diameter due to an increase in the Lankford value should also be considered when verifying the mechanism causing the excessive thinning of the outer diameter during microscale hollow sinking.

We previously reported that the final outer diameter of the microscale copper tubes was smaller than the die diameter during hollow sinking [10]. The mechanism causing this excessive thinning of the outer diameter was not elucidated in our previous work. This mechanism will be clarified by considering the size effect and plastic anisotropy, which have not been considered in conventional macroscale hollow sinking.

This study aims to verify the mechanism causing the excessive thinning of the outer diameter during microscale hollow sinking. In this study, the stainless-steel, copper, and aluminum alloy tubes, which have been reported to have different Lankford values [18], were drawn to investigate the effect of the Lankford value on the outer diameter reduction during hollow sinking. The metals selected were adopted from the face-centered cube (FCC) metals so that the effect of the crystal structure on the deformation behavior could be neglected. A metal's crystal structure seems to affect the deformation behavior of the drawn micro tube. Furthermore, the micro tubes were drawn at several drawing speed ratios to confirm that they yielded easily due to miniaturization. The measurement results of the outer diameter and the outer surface roughness of copper drawn tubes were already reported in our previous paper [10]. However, additional experiments were conducted to investigate the deformation behavior of the micro tubes in more detail using the stainless-steel and the aluminum alloy tubes. Therefore, the contents in this study were deeper, more detailed, and original.

2. Materials and Methods

2.1. Materials

Stainless-steel (SUS304), commercial purity copper (C1220), and aluminum alloy (A6063) tubes with an outer diameter D_0 of 1.50 mm and a wall thickness t_0 of 0.21 mm were used as starting materials. These materials were provided after drawing with a plug by Nippon Tokushukan MFG. Co., Ltd. (Osaka, Japan). The starting materials were annealed after plug drawing to relieve internal stresses. Tables 1 and 2 show the chemical compositions of the stainless-steel and the aluminum alloy tubes, which fall within the the reference range in accordance with Japanese Industrial Standards JIS-G4305 [19] and JIS-H4080 [20], respectively. The copper tubes used in this study were the same as those used in our previous study [10], which contained 99.96 mass percent copper, 0.02 mass percent phosphorus, and 0.02 mass percent other impurities. This chemical composition was in the reference range stated in Japanese Industrial Standards JIS-H3300 [21].

Table 1. Chemical composition of the stainless-steel (SUS304) tube used in this study (mass percentage).

C	Si	Mn	P	S	Ni	Cr	Fe
0.05	0.79	1.72	0.035	0.004	8.88	18.22	Bal.

Table 2. Chemical composition of the aluminum alloy (A6063) tube used in this study (mass percentage).

Si	Fe	Cu	Mn	Mg	Cr	Zn	Al
0.43	0.18	0.04	0.02	0.52	0.01	0.01	Bal.

2.2. Hollow Sinking

The starting materials were drawn without an inner tool using a draw-bench machine (Factory-Automation Electronics Inc., Osaka, Japan). A schematic illustration of the draw-bench machine is shown in Figure 1. The detailed drawing procedures and drawing conditions were previously reported [10]. Table 3 shows the drawing conditions. The die reduction R_e is defined by the outer diameter of the starting material D_0 and the die diameter D_{die}, as expressed in Equation (1).

$$R_e = 1 - \frac{D_{die}^2}{D_0^2} \tag{1}$$

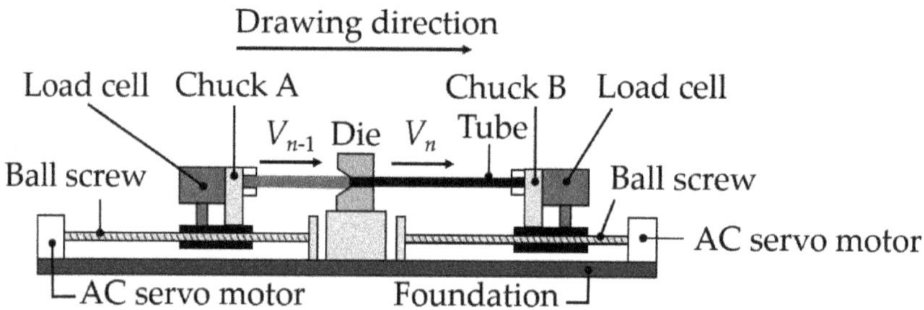

Figure 1. Schematic illustration of the draw-bench machine used in this study. The parameters V_{n-1} and V_n are the drawing speeds on the die's entrance and exit sides, respectively.

Table 3. Drawing conditions.

Die Reduction R_e	Drawing Speed Ratio β (= V_n/V_{n-1})		Die Half Angle $\theta°$
	1-chuck *1	2-chuck *2	
0.05	1.03	1.05, 1.10, 1.20	4
0.17	1.08	1.10, 1.20, 1.50	4
0.29	1.17	1.20, 1.50	4

*1: Only the tube on the exit side of the die was chucked. *2: Both the tubes on the entrance and exit sides of the die were chucked.

A load cell (PW6CC3MB 30 kg, Hottinger Brüel & Kjær GmbH, Darmstadt, Germany) was affixed to each chuck of the draw-bench machine. The load cell on the die's exit side measured the drawing force. The drawing force was calculated as the average load within 3% of the maximum load. The drawing stress was calculated by dividing of the drawing tension by the cross-sectional area of the tube on the die's exit side. The measurement method of the cross-sectional area is described in Section 2.4.

2.3. Tensile Test

Tensile tests of the starting materials and the drawn tubes were performed using a universal testing machine (Autograph AG 25 TB, Shimadzu Co., Kyoto, Japan), as shown in Figure 2. Tensile tests were performed to evaluate the followings: (1) the stress-strain curves of the starting materials and the drawn tubes, (2) the Lankford value of the starting materials at a maximum stroke, and (3) the unloading behavior of the drawn tube after reaching a drawing tension.

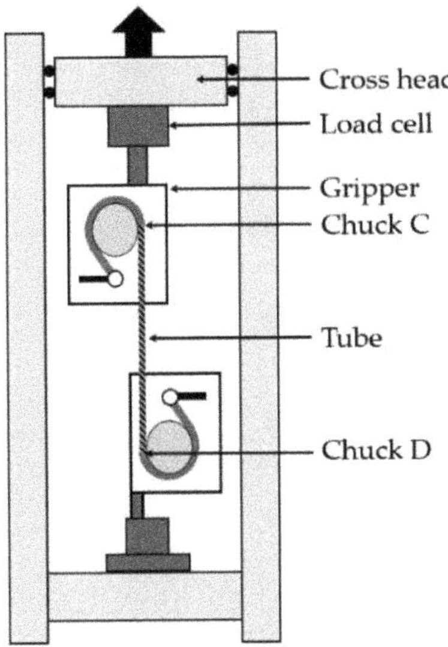

Figure 2. Schematic illustration of the method of tensile test used in this study.

The tube was grasped using a 5 kN grips set for fine wire (343-07529-01, Shimadzu Co., Kyoto, Japan). The distance between the chuck C and D was 100 mm before the test. The test speed was 5.0 mm/s. The test load was measured with a 5 kN load cell (SLBL-5kN, Shimadzu Co., Kyoto, Japan) affixed to the cross-head. The tensile test ended at a fracture point to obtain the stress-strain curves of the starting materials and the drawn tubes. The tensile tests of each material were conducted in triplicate. The nominal stress s was calculated by dividing the load by the initial cross-sectional area of the tube. The nominal strain e was calculated by dividing the stroke by the initial chuck distance. The true strain ε_{true} was calculated by converting the nominal strain to $\ln(e+1)$. The true stress σ_{true} was calculated by converting the nominal stress to $s(e+1)$.

The Lankford value during each maximum stroke was evaluated by conducting tensile tests at maximum strokes of 10, 20, 30, 40, and 50 mm. The Lankford value r was calculated using Equation (2). The average values were calculated, and the standard deviations were obtained.

$$r = \frac{\{\ln(D_{tensile}/D_0) + \ln(d_{tensile}/d_0)\}/2}{\ln(t_{tensile}/t_0)} \tag{2}$$

The parameters $D_{tensile}$ and $d_{tensile}$ are the outer diameter and inner diameter at each maximum stroke, respectively, as shown in Figure 3a. The method of measuring $D_{tensile}$ and $d_{tensile}$ is described in Section 2.4.

The loading-unloading tensile test was conducted to evaluate specimen unloading behavior. The drawn copper tube was loaded until reaching a load that corresponded to the drawing tension at a die reduction R_e of 0.17 and drawing speed ratio β of 1.10. The cross-head was moved until the load reached 2 N during unloading. The test speed during loading and unloading was 1.0 mm/s. An equivalent strain was obtained by evaluating the change in chuck distance during loading and unloading using a digital image correlate (DIC) software GOM Correlate^{TM} (2019, Gesellschaft für Optische Messtechnik GmbH, Braunschweig, Germany).

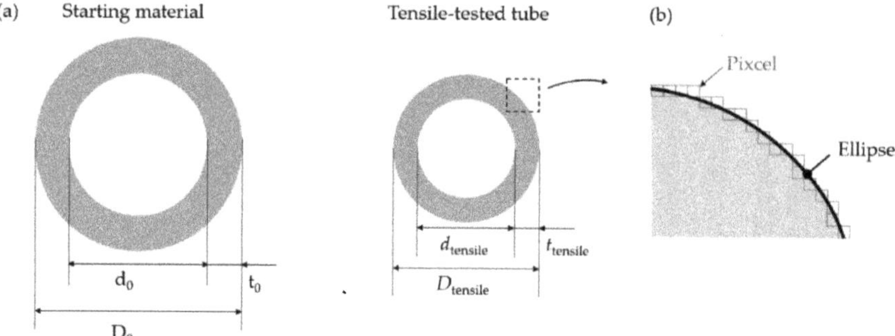

Figure 3. Measurement method of sample dimensions. (**a**) Dimension definition of the tube before and after the tensile test and (**b**) ellipse approximation of the outer circumference pixel.

Tensile test of the copper tube was also conducted using the draw-bench machine without the die to investigate the load measurement accuracy of the load cell affixed to the chuck of the draw-bench machine. The copper tube was chucked without setting the die. The initial chuck distance was 465 mm. The tube was stretched by moving the two chucks. The moving speeds of the chuck A and B were 3.3 mm/s and 5.0 mm/s, respectively. The stroke was calculated as the difference in the distance between the two chucks. The nominal stress was calculated by dividing the load by the initial cross-sectional area of the tube. The nominal strain was calculated by dividing the stroke by the initial chuck distance.

2.4. Dimension and Surface Quality Measurement

Cross sections of the drawn and tensile-tested tubes at each maximum stroke were polished and observed using a three-dimensional microscope (VHX-5000, Keyence, Miyagi, Japan). The outer diameter D and the inner diameter d of the tensile-tested and drawn tubes were measured by an image analysis as the following method. The pixels of the outer and inner circumference of the tube image were read using a proprietary programming language, MATLAB^{TM} (R2020a, Mathworks, Natick, MA, USA). A tube cross-section outline of the image became elliptical when the tube was cut diagonally. The minor axis of the ellipse was closest to the true diameter of the tubes. Therefore, ellipses were drawn by ellipse approximation of each circumference pixel, as shown in Figure 3b. The outer or inner diameter of the tube was measured as the minor axis of this ellipse. The wall thickness t was calculated as $(D - d)/2$. The cross-sectional area was calculated as $\pi t(D - t)$. The outer diameters of the drawn tube on the die's entrance and exit sides were measured using a micrometer (MDC-25M, Mitutoyo, Kanagawa, Japan) at four points. The average values were calculated, and the standard deviations were obtained.

Maps of the height of the uneven outer surface of the drawn tubes h were obtained using a laser microscope (VK9510, Keyence, Miyagi, Japan). The height from an arbitrary position of the uneven outer surface of the tube H was measured, excepting abnormally high or low value. The average values H_{ave} of the H were calculated. The height of the outer uneven surface of the tube h was calculated as the difference between the height from the arbitrary position H and the average height H_{ave} as $H - H_{ave}$.

2.5. Microstructural Observation

The number of crystal grains across the wall thickness t_0 of the starting materials was evaluated. The cross sections of the starting materials were mechanically polished after resin filling. Kikuchi patterns of the starting materials were obtained using a field emission scanning electron microscope (FE-SEM) (JSM-6500F, JEOL, Tokyo, Japan) with electron backscatter diffraction (EBSD) (OIM7, TSL, Kanagawa, Japan). The Inverse pole figure (IPF) maps were drawn, and the average grain size g was measured using an OIM Analysis^{TM} (ver.8, TSL, Kanagawa, Japan). A boundary with an orientation of 15° or more was defined as a grain boundary. The weighted mean of the grain area was calculated, and the average diameter of crystal grains was calculated as the grain size. The grain numbers across the wall thicknesses of the starting materials were calculated by dividing the wall thicknesses by the average grain size as t_0/g.

3. Results

3.1. Grain Number across Wall Thicknesses of Staring Materials

Figure 4 shows the IPF maps of the starting materials. The average grain size g of the stainless-steel, copper, and aluminum alloy tubes were 9 μm, 17 μm, and 71 μm, respectively. Therefore, the grain numbers across the wall thicknesses of starting materials t_0/g of the stainless-steel, copper, and aluminum alloy tubes were 23, 12, and 3, respectively.

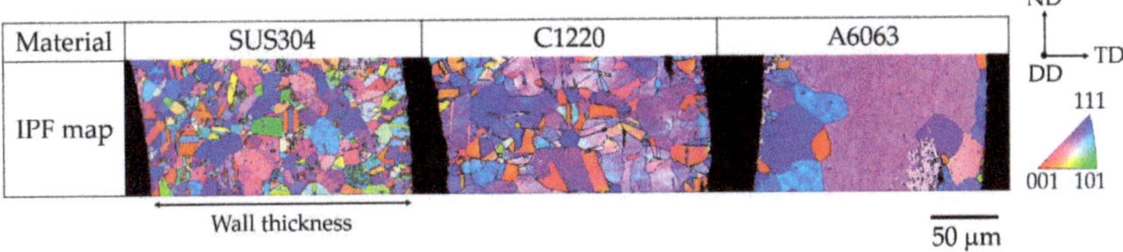

Figure 4. Inverse pole figure (IPF) maps of the starting materials. The symbols DD, TD, and ND indicate drawing direction, transversal direction, and normal direction, respectively. The grain boundaries were drawn on as black lines on IPF maps, using the OIM Analysis[TM].

3.2. Lankford Value of Starting Materials

Figures 5 and 6 show the stress-strain curves and Lankford values r of each starting material, respectively. The true strain $\varepsilon_{\text{true}}$ in Figure 6 corresponds to that in Figure 5b. A drop in stress due to necking was not observed for all materials, as shown in Figure 5a. Necking was observed only on the fractured surface of the copper tube. It is considered that necking occurred immediately before the fracture. Therefore, the strain from necking to fracture was negligible. The measured fracture strain of the aluminum alloy tube was 0.31. Therefore, the Lankford values r of the aluminum alloy tube at the true strain of 0.34 and 0.41 could not be measured. The Lankford value decreased as the true strain increased for each material. The Lankford values decreased in the order of the aluminum alloy, the copper, and the stainless-steel tube at each true strain.

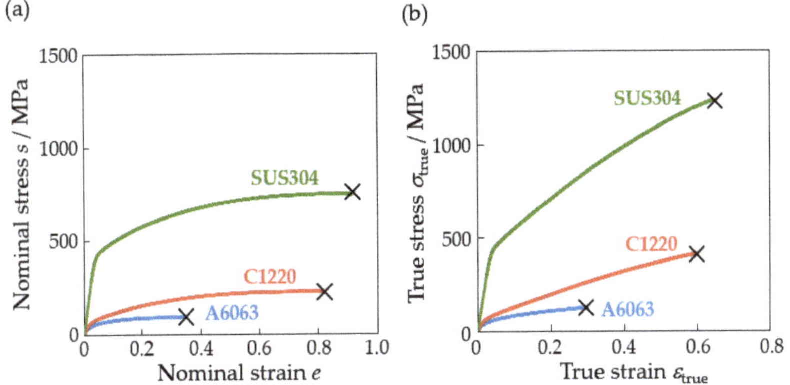

Figure 5. Results of the tensile test. (**a**) Nominal stress—nominal strain curves and (**b**) true stress—true strain curves. The symbol × indicates the fracture point.

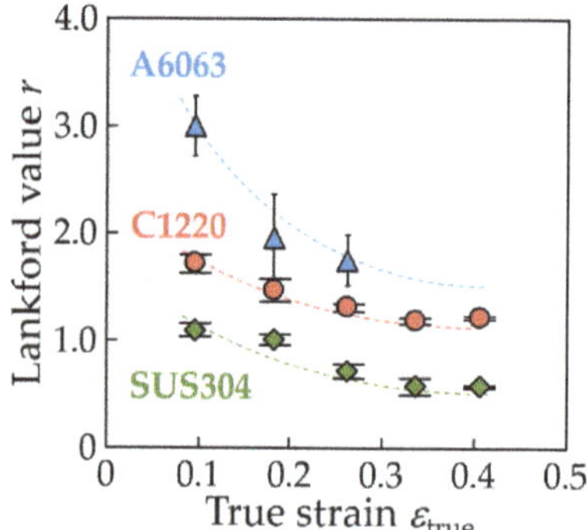

Figure 6. Lankford value against true strain obtained by the tensile test of each material. The dotted lines indicate the eye guide.

3.3. Dimension of Drawn Tubes

Figure 7 shows the cross sections of the drawn tubes of the selected materials at a die reduction of 0.17 (die diameter D_{die} = 1.37 mm). The results observed on the copper tube were already reported in our previous paper [10]. The dotted circles in the image area are equivalent to the die diameter. For each material, all the cross sections were smaller than the die's exit hole. The cross section became smaller than the die's exit hole as the drawing speed ratio increased. Figure 8 shows the measurement results of the outer diameter of the drawn tubes on the die's exit side for each drawing condition. The error bars indicate the standard deviations. The dotted lines indicate the die diameter. The outer diameter was smaller than the die diameter for all drawing conditions. The outer diameter became smaller than the die diameter as the drawing speed increased.

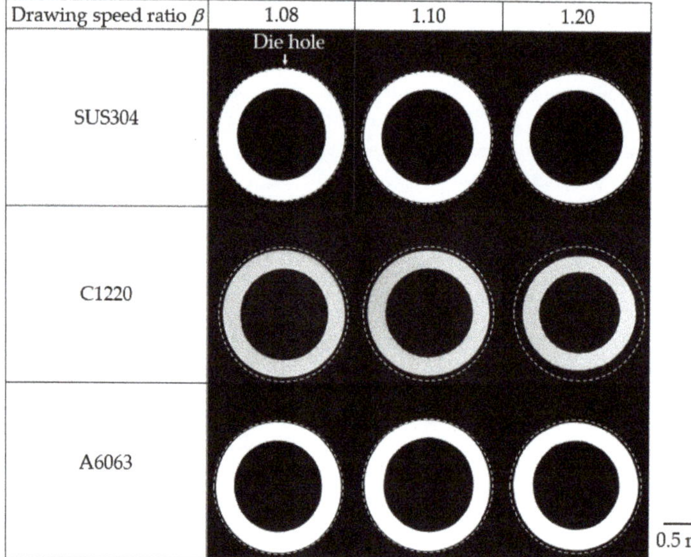

Figure 7. Cross sections of the drawn tubes at a die reduction R_e of 0.17 for each material. The diameter of the dotted circle was drawn to show the size of the exit hole of the die. The results of the copper tubes were already shown in our previous paper [10].

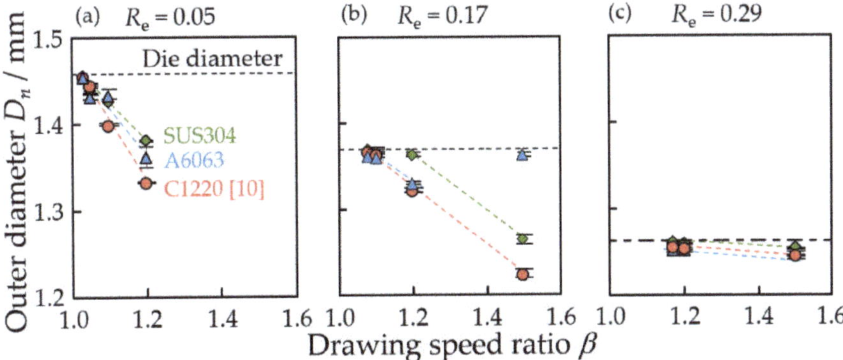

Figure 8. Outer diameters of the drawn tubes on the die exit side. (**a**) Die reduction R_e = 0.05, (**b**) R_e = 0.17, and (**c**) R_e = 0.29. The error bars indicate the standard deviations. The dotted lines indicate the exit hole diameter of the die. The results of the copper tubes were already shown in our previous paper [10].

3.4. Outer Surface Quality of Drawn Tubes

Figure 9 shows the height maps of the outer surface of the drawn tubes. Since the outer diameter was estimated to be much smaller than the die diameter before the tube drawing test, tube drawing at a drawing speed ratio of 1.50 was not performed at a die reduction of 0.05. Asterisks indicate that the tube drawing was not performed for this reason. The minimum drawing speed ratio for each die reduction is shown in Table 3. The symbol ** indicates drawing experiments that were impossible to perform because the drawing speed ratio was less than the minimum value. The number of areas with regions of particularly high (red) or low (blue) heights increased as the drawing speed ratio increased for each material. The proportion of low-height and the high-height parts decreased in the order of the copper, the aluminum alloy, and the stainless-steel tube. Therefore, the outer surfaces of the materials were rougher in the order of the copper, the aluminum alloy, and the stainless-steel tube for drawing conditions.

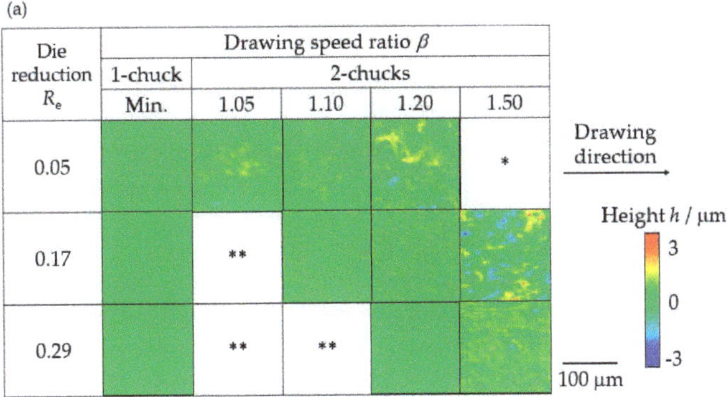

Figure 9. *Cont.*

(b)

Die reduction R_e	Drawing speed ratio β				
	1-chuck	2-chucks			
	Min.	1.05	1.10	1.20	1.50
0.05					*
0.17		**			
0.29		**	**		

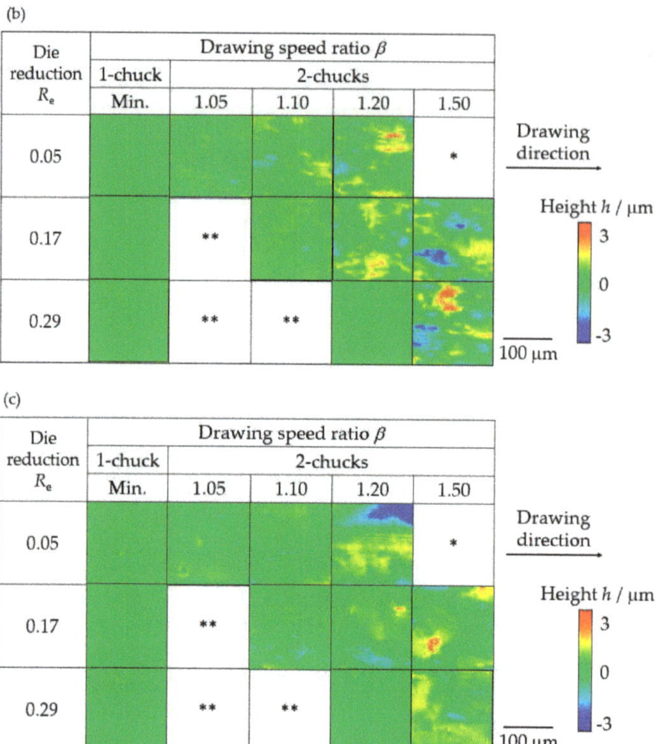

(c)

Die reduction R_e	Drawing speed ratio β				
	1-chuck	2-chucks			
	Min.	1.05	1.10	1.20	1.50
0.05					*
0.17		**			
0.29		**	**		

Figure 9. Height maps of the outer surface of the drawn tubes. (**a**) Stainless-steel, (**b**) copper [10], and (**c**) aluminum alloy tubes. 1-chuck implies that only the tube on the exit side was chucked. 2-chucks means that both the tubes on the die entrance and exit sides were chucked. The symbol * indicates that the drawing experiment was not performed. The symbol ** indicates that the drawing experiment was impossible because the drawing speed ratio is less than the minimum value.

3.5. Drawing Stress during Drawing

The mechanism causing the excessive thinning of the diameter was discussed based on the drawing stress, calculated from the load measured by the load cell affixed to the chucks of the draw-bench machine. Therefore, the load measurement accuracy of the draw-bench machine was investigated. The load measured by each load cell affixed to chucks A and B should match when a tube is simply stretched. It was confirmed that the two load cells affixed to chucks A and B measured an equivalent load, as shown in Figure 10. The load measurement accuracy of the draw-bench machine was evaluated by comparing the load obtained by the universal testing machine and the draw-bench machine. The load of the chuck A was applied for comparison. Figure 11 shows a comparison of the load measured by the universal testing machine and the draw-bench machine. In the true strain range larger than 0.4, the load of the draw-bench machine was up to 2% smaller than that of the universal testing machine. However, the drawing tension under most of the experimental conditions in this study was in a strain range smaller than 0.4. Therefore, the load measurement accuracy of the draw bench was allowed when compared to the universal testing machine.

Figure 12 shows measurement results of the copper tube's load at the die reduction R_e of 0.17 and the drawing speed ratio of 1.10. The drawing force was calculated as the average load within 3% of the maximum. Figure 13 shows the measured drawing stress at the die reduction of 0.17 and 0.29 for each material. The tubes were drawn without contacting the die during drawing under an increased drawing speed ratio when the die reduction was 0.05. Therefore, the results of the drawing stress at the die reduction R_e of 0.05 are not shown. The drawing stress increased as the drawing speed ratio increased for each material. The drawing stress was large in the order of the stainless-steel, the copper, and the aluminum alloy tubes. This order corresponded to the fracture stress of the stress-strain curves in Figure 5.

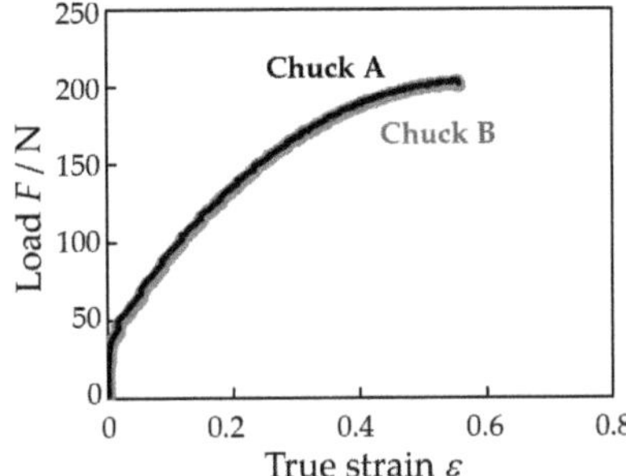

Figure 10. Measurement results of the load obtained by the tensile test of the copper tube using the draw-bench machine without the die. Chucks A and B indicate the chuck of the draw-bench machine on the die's entrance and exit sides, respectively.

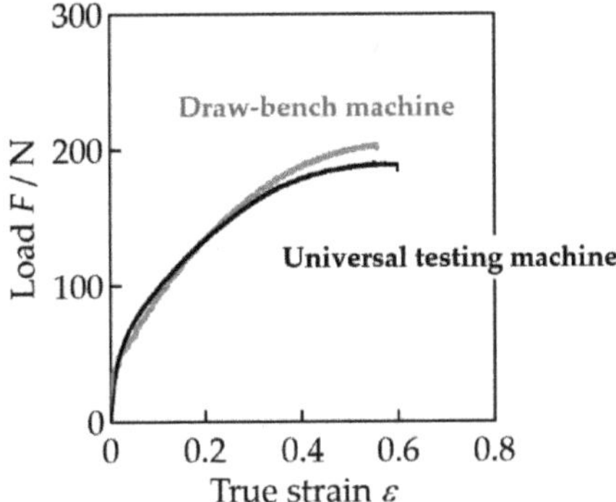

Figure 11. Comparison of the load obtained by the tensile test using the universal testing machine and the draw-bench machine without the die.

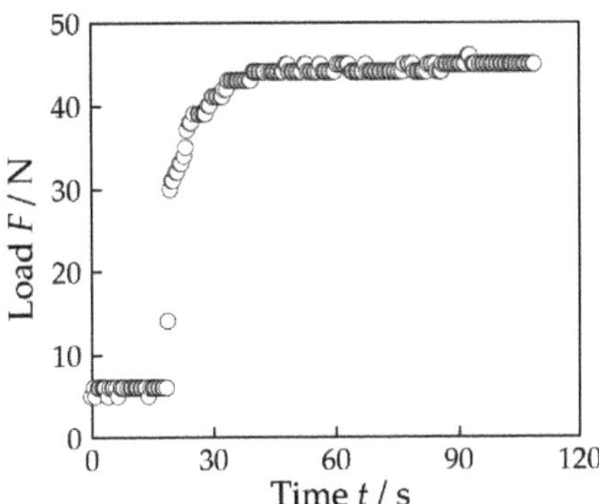

Figure 12. Load during drawing of the copper tube at the die reduction R_e of 0.17 and the drawing speed ratio β of 1.10.

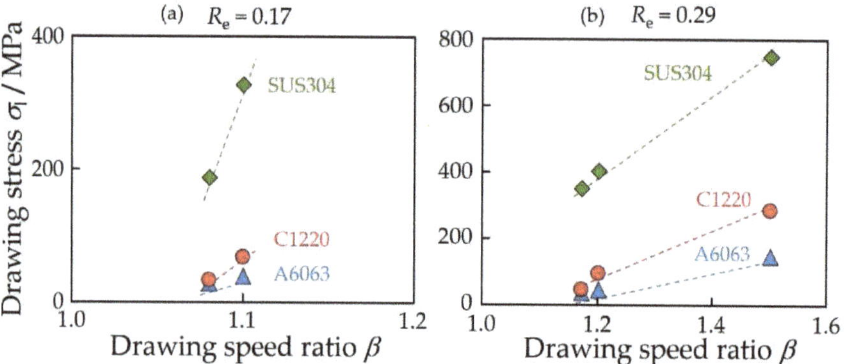

Figure 13. Drawing stress during drawing for each material. (**a**) Die reduction was 0.17 and (**b**) 0.29. The dotted lines indicate the eye guide.

4. Discussion

4.1. Conditions to Prevent Non Contact between the Tube and Die

We have previously reported that the copper tubes contacted the die under the conditions where the outer diameter on the die's entrance side was larger than the die diameter, as shown in the gray area of Figure 14b [10]. Therefore, the drawing speed ratio should be set low so that the die diameter increases, which prevents non contact between the tube and the die. Figure 14a,c show the outer diameter on the die's entrance side of stainless-steel and aluminum alloy drawn tubes, respectively. Similar to our previous report [10], the tube contacted the die at a small die diameter and a low drawing speed ratio. Only the aluminum tube contacted the die at a die diameter D_{die} of 1.37 mm and a drawing speed ratio β of 1.50. In the following sections, the excessive thinning of the outer diameter was discussed only under the conditions where the tube contacted the die.

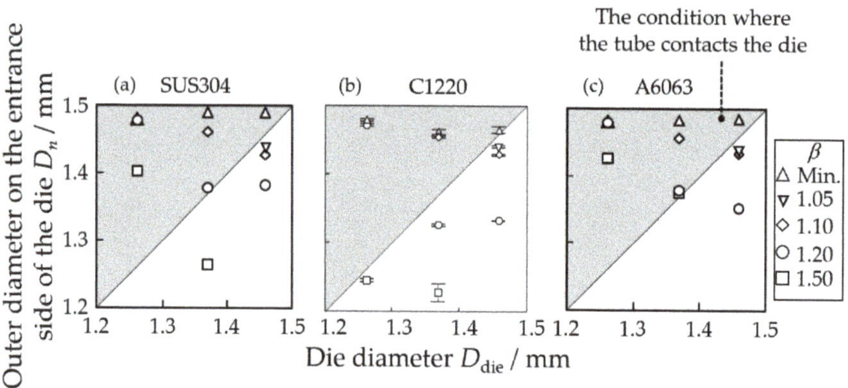

Figure 14. Outer diameter of the drawn tubes on the die entrance side for each material. (**a**) Stainless-steel, (**b**) copper, and (**c**) aluminum alloy tubes. The results of copper tubes were published in our previous paper [10].

4.2. Mechanism of the Excessive Thinning of the Outer Diameter during and after Drawing

4.2.1. Deformation Behavior of Micro Tube during Drawing and Unloading

We have previously reported that the outer diameter of the micro tube decreased from the die approach end [10]. Therefore, the excessive thinning of the outer diameter was considered as a uniaxial tensile deformation starting from the die approach end. To investigate this deformation behavior, the magnitude of the drawing stress in the stress-strain curve of the drawn tubes was evaluated. Figure 15a shows the stress-strain curve of the drawn stainless-steel tube at a die reduction R_e of 0.29 and a drawing speed ratio β of 1.20. The dotted line indicates the drawing stress σ_l during drawing. The state where the outer diameter matched the die diameter after passing through the die approach corresponds to the origin in Figure 15a. Generally, a bulk metal that has been deformed under a load in the elastic range will recover to its original shape as soon as the load is removed. In this study, the drawing stress σ_l was in the macroscopic elastic region of the stress-strain curve. However, the outer diameter finally decreased excessively from the die approach end as the plastic strain. Therefore, it is considered that the excessive thinning of the outer diameter was caused by microscopic yielding of the micro tube due to the small number of crystal grains across the wall thickness, as shown in Figure 15b. Bulk metal, where many crystal grains exist across the thickness, yields microscopically even under macroscopically elastic deformation behavior [22]. Therefore, it is considered that the micro tube used in this study easily yielded microscopically because of the small number of crystal grains across the wall thickness. Furthermore, the apparent elastic modulus E' of the micro stainless-steel tube in Figure 15a was 24 GPa, which was much smaller than that of the reference value of bulk stainless-steel (204 GPa [23]). The reason for the decrease in the apparent elastic modulus due to the miniaturization reported elsewhere has not been clarified [24]. However, it is considered that this decrease in the apparent elastic modulus was also caused by microscopic yielding. The slope during loading in the macroscopic elastic range seems to decrease because of easy yielding for a certain drawing stress. The measurement accuracy of the true stress using the universal testing machine was confirmed in Section 3.5. Therefore, the apparent elastic modulus in Figure 5b seems to be appropriate.

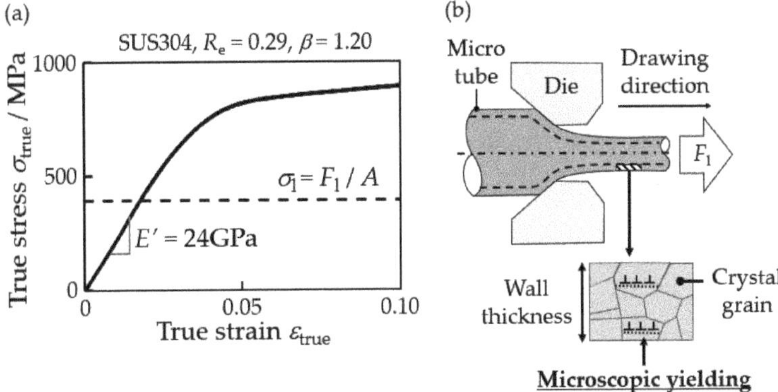

Figure 15. Measurement results of the stress-strain curve and drawing stress during drawing. (a) Stress-strain curve of the drawn stainless-steel tube at the die reduction R_e of 0.29 and the drawing speed ratio β of 1.20. The dotted line indicates the drawing stress σ_l. The parameters F_l and A are the drawing tension and the cross-sectional area of the drawn tube, respectively. The parameter E' is the apparent elastic modulus of the stress-strain curve. (b) Schematic illustration of the microscopic yielding of the drawn micro tube during drawing.

Figure 16 shows the result of the loading-unloading tensile test of the drawn copper tube at a die reduction R_e of 0.17 and a drawing speed ratio β of 1.10. Figure 16b shows the equivalent strain distribution during loading. The strain distribution between the chucks was uniform. Therefore, it is considered that the true strain can be evaluated by measuring the change in the chuck distance. The equivalent strain was also obtained using DIC by measuring the chuck distance. The total strain ε_{total} and the unloading strain $\Delta\varepsilon_{unload}$ obtained using the universal testing machine and the DIC matched with errors of 4.3% and 9.2%, respectively, as shown in Figure 16c. The measurement accuracy of the true strain using the universal testing machine was almost equivalent to that of the DIC in this study. The measurement accuracy of the true stress using the universal testing machine was confirmed in Section 3.5. Therefore, the apparent elastic modulus in Figure 5b seems to be appropriate.

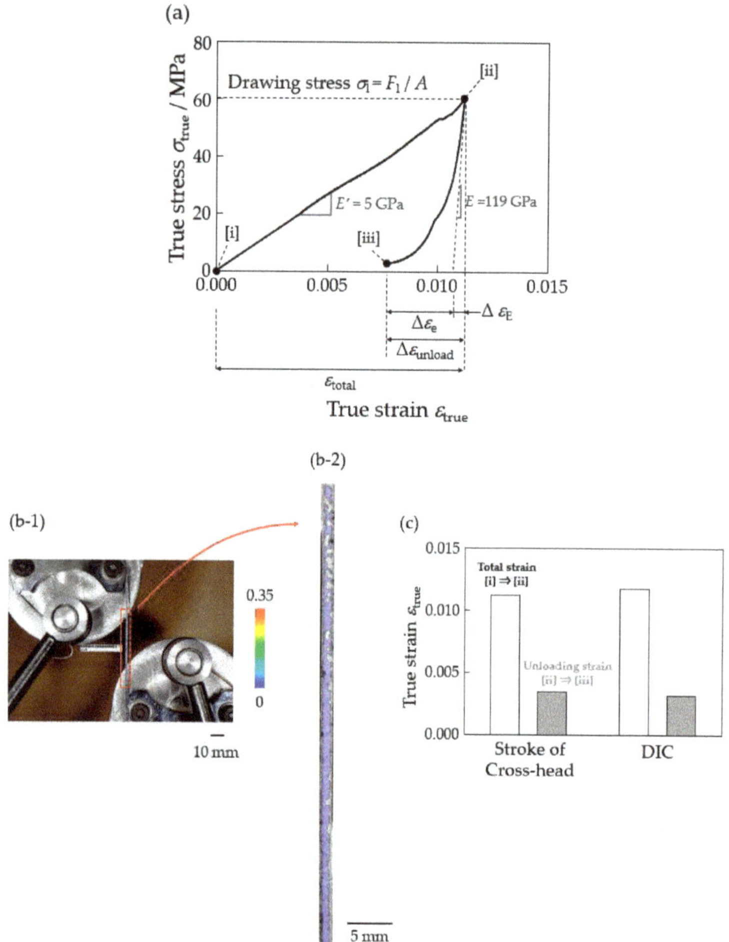

Figure 16. Results of the loading-unloading tensile test of the drawn copper tube at the die reduction R_e of 0.17 and the drawing speed ratio β of 1.10. (**a**) Loading-unloading curve. The parameters $\Delta \varepsilon_E$ and $\Delta \varepsilon_{total}$ are the elastic strain and the total strain, respectively. The parameter E' is the apparent elastic modulus during loading. The symbols [i], [ii], and [iii] indicate the origin, the end point of loading, and the end point of unloading, respectively, (**b-1**) distribution of the equivalent strain, (**b-2**) magnified view of (**b-1**), and (**c**) comparison of the true strain obtained by the universal testing machine and the DIC. The symbols [i]–[iii] coresspond to Figure 16a.

The true stress ε_{true} increased during loading until reaching the drawing stress σ_l, as shown in Figure 16a. The apparent elastic modulus during loading E' was 5 GPa, which was much smaller than that of the reference value of the bulk copper (119 GPa [23]). This decrease in the apparent elastic modulus due to miniaturization was also shown in the stress-strain curve of the starting material as shown in Figure 15. This decrease in the apparent elastic modulus was considred to be a result of microscopic yielding. The slope during loading seems to decrease because of easy yielding for a certain drawing stress. The true stress ε_{true} dropped slightly at a true strain of approximately 0.010 during loading. The reason for this stress drop is unclear. However, the true strain during this dropping was 0.7% against the total strain ε_{total}. Therefore, it is considered that this stress drop was negligible.

The true strain did not recover completely during unloading. The unloading strain/total strain was 0.31. This result also indicates that the micro tube yielded microscopically during loading. The outer diameter seems to approach the die diameter as the unloading strain $\Delta\varepsilon_{unload}$ increases. It is apparent that the unloading behavior depends on the elastic recovery outside the microscopically yielded region. Therefore, the unloading strain at the initial stage of the unloading behavior seems to depend on the elastic strain $\Delta\varepsilon_E$, which was calculated by dividing the drawing stress σ_l by the bulk elastic modulus E of 119 GPa [23]. However, the unloading strain $\Delta\varepsilon_{unload}$ was significantly larger than the elastic strain $\Delta\varepsilon_E$. Therefore, it is considered that the outer diameter of the micro tube increased more than the linear elastic strain $\Delta\varepsilon_E$ during unloading. The difference between the unloading strain and the elastic strain is defined as the excessive elastic strain $\Delta\varepsilon_e$, which is discussed in Section 4.2.3.

According to the above discussion, the following two deformation behaviors should be investigated to clarify the mechanism that causes the excessive thinning of the outer diameter: (1) the excessive thinning of the outer diameter due to the microscopic yielding during drawing and (2) the unloading behavior caused by elastic recovery outside the microscopically yielded region during unloading, which determines the final outer diameter. Therefore, the followings were investigated: (1) the relationship between the total strain ε_{total} and the outer diameter during drawing D_{total} and (2) the relationship between the unloading strain $\Delta\varepsilon_{unload}$ and the final outer diameter D_n. The detail procedures are shown in Figure 17. The outer diameter during drawing D_{total} and the unloading strain $\Delta\varepsilon_{unload}$ could not be measured. Therefore, the outer diameter during drawing D_{total} was calculated by using the Lankford value, which indicated the plastic anisotropy, and the total strain ε_{total}. The unloading strain $\Delta\varepsilon_{unload}$ was calculated by using the plastic strain ε_p and the total strain ε_{total}.

The tensile residual stress, which is generated during drawing [25], seems to hinder the unloading behavior, because the force directions of the unloading behavior and the tensile residual stress are opposing. Tensile residual stress is not generated during uniaxial tensile deformation. Therefore, the difference in the unloading behavior between the tensile test and the drawing test is discussed based on the tensile residual stress in Section 4.2.3.

(a)

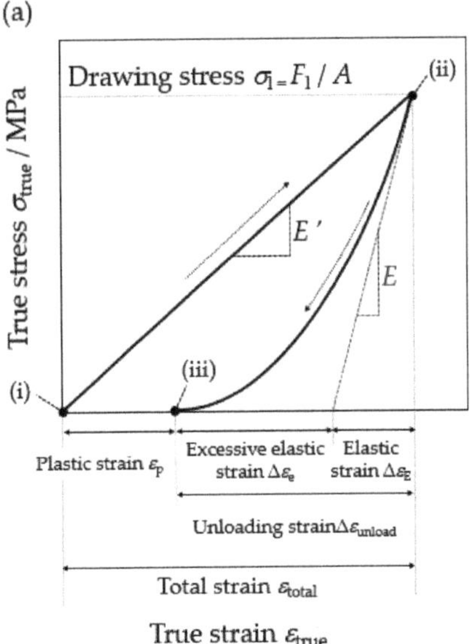

Figure 17. *Cont.*

(b)

	(0) Before drawing		(i) After passing through die approach	(ii) During drawing	(iii) After unloading
	(0-1)	(0-2)			
Schema	Micro tube Die DD	BT	Die approach	F_1	
Outer diameter	D_0 (measured value)		D_{die}	D_{total}	D_n (measured value)
Wall thickness	t_0 (measured value)		t_{die}	t_{total}	t_n (measured value)
Length	l_0		l_{die}	l_{total}	l_n

Figure 17. Road map for clarifying the mechanism causing the excessive thinning of the outer diameter. (**a**) Each strain in the stress-strain curve of the drawn tube. The parameters E' and E are the apparent elastic modulus and the elastic modulus of the bulk metal, respectively, (**b**) dimensions for each state. The symbol DD indicates the drawing direction. The parameters F_1 and F_{BT} are the drawing tension and the back tension, respectively.

4.2.2. Excessive Thinning of the Outer Diameter during Drawing

The relationship between the total strain ε_{total} and the outer diameter during drawing D_{total} was investigated. The deformation of the micro tube on the die's entrance side due to the back tension (state (0-2) shown in Figure 17) was small when compared to the main deformation (state (0-1)→(ii)). Therefore, this deformation was neglected. It is assumed that the outer diameter decreased during drawing from the state where the outer diameter matched with the die diameter D_{die}, as (i)→(ii) shown in Figure 17. The outer diameter during drawing D_{total} was calculated using the die diameter D_{die}, the wall thickness after passing through the die approach t_{die}, the Lankford value r in deformation during drawing, and the total strain ε_{total}. The wall thickness after passing through the die approach t_{die} and the Lankford value r were calculated using Equations (3) and (4), respectively. The detailed deviations of Equations (3) and (4) are shown in the Appendix A.

$$t_{die} = \frac{1}{2}\left(D_{die} + \sqrt{D_{die} - \frac{4t_0(d_0 - t_0)}{\beta}} \right)$$

(3)

$$r = \frac{\ln(D_n/D_{die})}{\ln(t_n/t_{die})}$$

(4)

Equation (5) indicates the total strain during drawing. By substituting Equation (4) into Equation (5) to eliminate the wall thickness t_{total}, Equation (6) is obtained.

$$\varepsilon_{total} = \ln\frac{l_{total}}{l_{die}} = \ln\frac{t_{die}(D_{die} - t_{die})}{t_{total}(D_{total} - t_{total})}$$

(5)

$$r = \frac{\ln(D_{total}/D_{die})}{\ln\dfrac{D_{total} \pm \sqrt{D_{total} - \dfrac{D_{die} - t_{die}}{\exp(\varepsilon_{total})}}}{2t_{die}}}$$

(6)

The outer diameter during drawing D_{total} was calculated using Equation (6). The excessive thinning of the outer diameter during drawing η' was calculated using Equation (7). Furthermore, the excessive thinning of the outer diameter after drawing η was calculated using Equation (8). The measurement results of the final outer diameter D_n are already shown in Figure 8.

$$\eta' = \frac{D_{die} - D_{total}}{D_{die}} \times 100$$

(7)

$$\eta = \frac{D_{\text{die}} - D_n}{D_{\text{die}}} \times 100 \tag{8}$$

Figure 18a shows the theoretical relationship between the total strain $\varepsilon_{\text{total}}$ and the excessive thinning of the outer diameter during drawing η'. η' increases as the total strain $\varepsilon_{\text{total}}$ increases, and further grows during drawing as the Lankford value r increases. η' is greater than zero at any Lankford value. Therefore, it is considered that the outer diameter always becomes much smaller than the die diameter during drawing at any Lankford value r. Figure 18b shows η' in this study. Figure 18c–e show the Lankford value of each η' in Figure 18a. η' was obtained by substituting t_{die}, $\varepsilon_{\text{total}}$, and D_{die} into Equation (6) for each drawing condition. The theoretical excessive thinning of the outer diameter during drawing η' was larger in the order of the aluminum alloy, the copper, and the stainless-steel tube corresponding to the order of the Lankford values r in Figure 6. Therefore, the relationship of the magnitude between the theoretical values for each material seems to be appropriate, and the excessive thinning of the outer diameter during drawing η' increased as the Lankford value grew for each material.

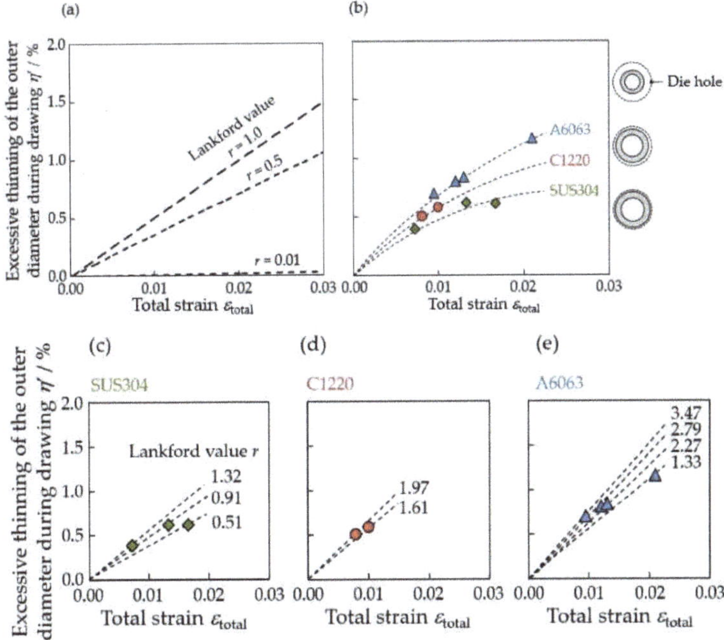

Figure 18. Theoretical relationship between the Lankford values r and the excessive thinning of the outer diameter during drawing η'. (**a**) The excessive thinning of the outer diameter at the Lankford values r of 0.01, 0.5, and 1.0. (**b**) The excessive thinning of the outer diameter in this study of all materials in this study, including (**c**) stainless-steel, (**d**) copper, and (**e**) aluminum alloy tubes. The dotted lines or curves in (**b**–**d**) indicate the eye guide.

4.2.3. Excessive Thinning of the Outer Diameter after Drawing

The relationship between the unloading strain $\Delta\varepsilon_{\text{unload}}$ and the final outer diameter D_n (state(iii) in Figure 17a,b) was investigated. The unloading strain $\Delta\varepsilon_{\text{unload}}$ was calculated as the difference between the total strain $\varepsilon_{\text{total}}$ and plastic strain ε_p, which is required for deformation to the final dimensions described by Equations (9) and (10).

$$\varepsilon_p = \ln\frac{l_n}{l_{\text{die}}} = \ln\frac{t_{\text{die}}(D_{\text{die}} - t_{\text{die}})}{t_n(D_n - t_n)} \tag{9}$$

$$\Delta\varepsilon_{\text{unload}} = \varepsilon_{\text{total}} - \varepsilon_{\text{p}} \tag{10}$$

Figure 19 shows the relationship between the unloading strain $\Delta\varepsilon_{\text{unload}}$ and the final excessive thinning of the outer diameter η. The final excessive thinning of the outer diameter η decreased when the unloading strain $\Delta\varepsilon_{\text{unload}}$ increased. Therefore, the final excessive thinning of the outer diameter decreased as the drawn tube recovered elastically during unloading.

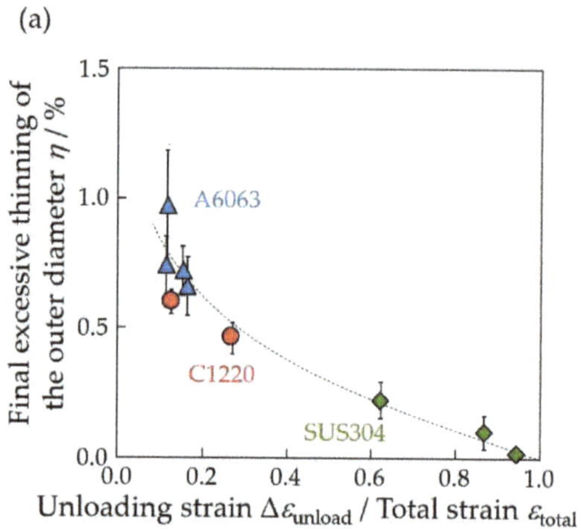

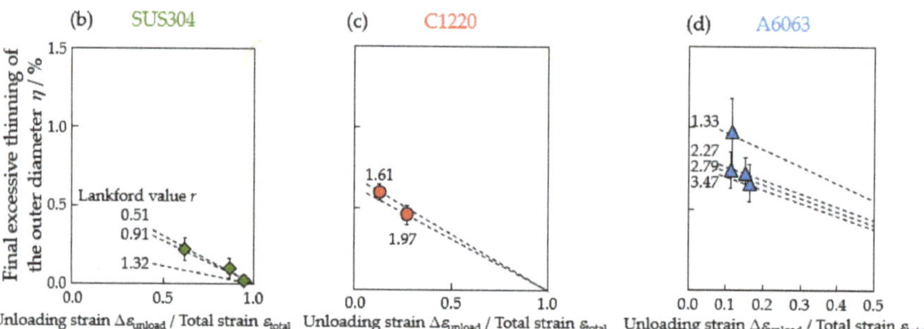

Figure 19. Relationship between the unloading strain $\Delta\varepsilon_{\text{unload}}$ and the final excessive thinning of the outer diameter η, (**a**) all materials, (**b**) stainless-steel, (**c**) copper, and (**d**) aluminum alloy tubes. The dotted lines indicate the eye guide.

The physical meaning of the unloading strain $\Delta\varepsilon_{\text{unload}}$ is discussed as follows. The elastic strain $\Delta\varepsilon_{\text{E}}$, which depends on the elastic modulus of the bulk metal, was significantly smaller against the unloading strain of the loading-unloading tensile test, as shown in Figure 16. The elastic strain $\Delta\varepsilon_{\text{E}}$ in hollow sinking was calculated using Equation (11).

$$\Delta\varepsilon_{\text{E}} = \sigma_{\text{l}}/E \tag{11}$$

The parameter E is the elastic modulus of the bulk metal. The reference value of the elastic modulus of each bulk metal is substituted into Equation (11). The reference values of stainless-steel, copper, and aluminum alloy are 204 GPa [23], 119 GPa [23], and 69 GPa [26], respectively. The elastic strain value in the range of 0.0005 to 0.002 was significantly smaller against the unloading strain of 0.002 to 0.015. Therefore, the outer diameter of the micro tube approached the die diameter due to the unloading strain, which was larger than the elastic strain, during unloading. The excessive elastic strain $\Delta\varepsilon_e$, which was the difference between the unload strain $\Delta\varepsilon_{unload}$ and the elastic strain $\Delta\varepsilon_E$, was calculated using Equation (12).

$$\Delta\varepsilon_e = \Delta\varepsilon_{unload} - \Delta\varepsilon_E \tag{12}$$

Figure 20a,b show the relationship between the excessive elastic strain $\Delta\varepsilon_e$ and the apparent elastic modulus E', and the drawing stress σ_l. The excessive elastic strain $\Delta\varepsilon_e$ grew when the apparent elastic modulus increased, or when the drawing stress decreased. The final outer diameter approached the die diameter as the drawing speed ratio decreased in Figure 8. Furthermore, the final outer diameter D_n fell in the order of the stainless-steel, the copper, and the aluminum alloy tube as shown in Figure 8. Figure 21 schematically shows the physical meaning of the excessive elastic strain. The micro tube recovered elastically to a degree greater than the elastic strain during unloading. Therefore, it is considered that the dislocations generated by microscopic yielding that occurred during drawing, disappeared partially during unloading. This phenomenon seems to be equivalent to the Bauschinger effect. It is considered that the dislocation density remaining after unloading was small under the conditions where few dislocations were generated during drawing such as with a high apparent elastic modulus, or low drawing stress. Therefore, the excessive elastic strain increased when the apparent elastic modulus grew, or the drawing stress decreased.

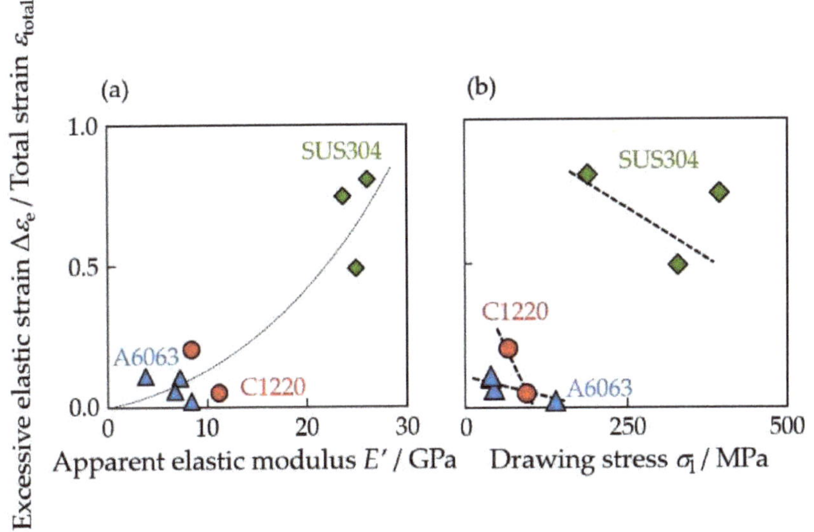

Figure 20. Relationship between the excessive elastic strain $\Delta\varepsilon_e$ and (**a**) the apparent elastic modulus E', (**b**) drawing stress σ_l. The dotted curve and line indicate the eye guide.

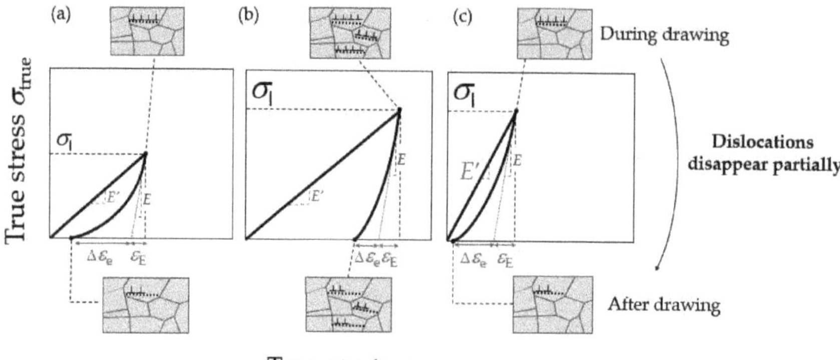

Figure 21. Schematic illustration to explain the physical meaning of the excessive elastic strain $\Delta\varepsilon_e$. The excessive elastic strain $\Delta\varepsilon_e$ (**a**) at small drawing stress σ_1 and small apparent elastic modulus E', (**b**) at large drawing stress σ_1 and small apparent elastic modulus E', and (**c**) at large drawing stress σ_1 and large apparent elastic modulus E'. The parameters E and $\Delta\varepsilon_E$ are the elastic modulus of the bulk metal and the elastic strain, respectively.

The unloading strain/total strain during the tensile test of the copper tube in Figure 16 was larger than that of the drawing experiment in Figure 19. Generally, the tensile residual stress is generated in the longitudinal direction during drawing [25]. Since the direction of the forces associated with the unloading behavior and the tensile residual stress are opposite, the tensile residual stress is considered to hinder the unloading behavior after drawing. Therefore, the amount of strain recovery during the tensile test was larger than that of the drawing test. A significant difference in tensile residual stress occurs when the die half angle is changed for the tube drawing process. Tube drawing was performed in this case using only one die half angle. Therefore, the effect of the tensile residual stress on the unloading strain is negligible. Investigation of the effect of the tensile residual stress on the unloading strain remains a subject of future research.

An approximation of the unloading behavior was performed to investigate the effect of the apparent elastic modulus on the unloading behavior. Figure 22 shows the approximation method for the unloading behavior.

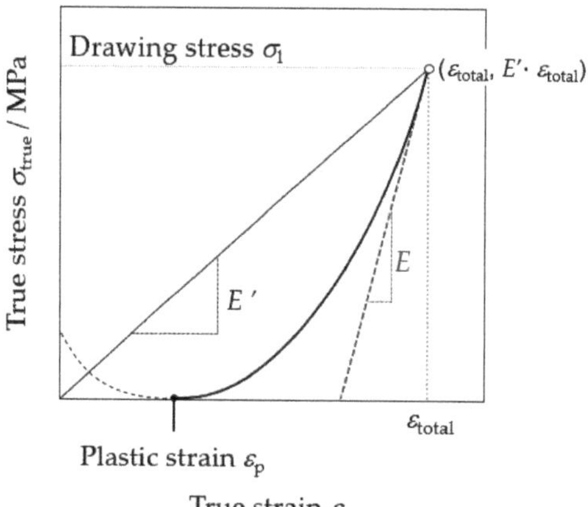

Figure 22. Illustration of the approximated unloading behavior.

The unloading behavior was approximated by a quadratic function, as shown in Equation (13). The parameter a is constant. By considering that the slope at the point $(\varepsilon_{total}, E' \cdot \varepsilon_{total})$ is E, Equation (14) is obtained.

$$\sigma_{true} = a\left(\varepsilon - \varepsilon_p\right)^2 \tag{13}$$

$$a = \frac{E}{2\left(\varepsilon_{total} - \varepsilon_p\right)} \tag{14}$$

The plastic strain ε_p is obtained by substituting ε_{total} and $E' \cdot \varepsilon_{total}$ into ε and σ_{true} of Equation (15), respectively, as shown in Equation (15).

$$\varepsilon_p = \varepsilon_{total} - \frac{2E' \cdot \varepsilon_{total}}{E} \tag{15}$$

Lastly, substituting the plastic strain into Equation (15), Equation (16) is obtained as follows.

$$\sigma_{true} = \varepsilon_{total} - \frac{E^2\left(\varepsilon - \varepsilon_{total} + \frac{2E' \cdot \varepsilon_{total}}{E}\right)^2}{4E' \cdot \varepsilon_{total}} \tag{16}$$

Figure 23 compares the actual plastic strain calculated by Equation (9) and calculated by Equation (15). The experimental values of the total strain ε_{total}, the bulk elastic modulus E, and the apparent elastic modulus E' are substituted into Equation (15). The plastic strain obtained by Equation (15) generally agrees with the value obtained from Equation (9). Therefore, the unloading behavior can be expressed by a quadratic approximation. Figure 24 shows the unloading behavior obtained by Equation (16) at the bulk elastic modulus E and the apparent elastic modulus E' of 190 GPa and 95 GPa, respectively. The experimental value of the total strain ε_{total} is substituted into Equation (16). The strain completely recovered after unloading. Therefore, it is considered that the dislocations caused by the microscopic yielding disappear completely during unloading when the apparent elastic modulus is larger than a threshold value.

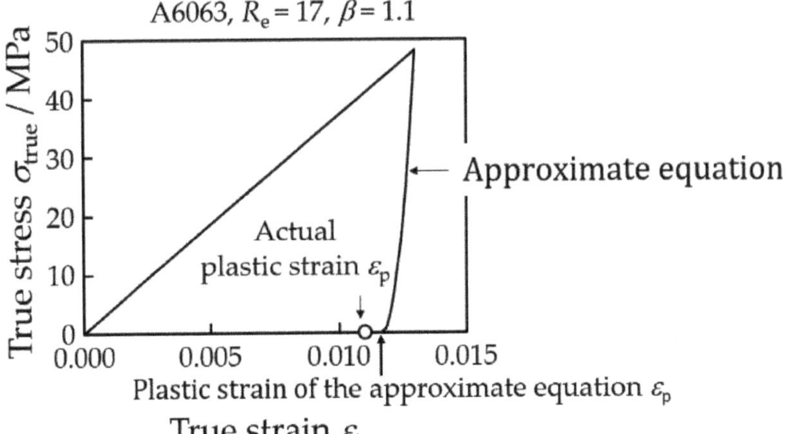

Figure 23. Comparison of the actual plastic strain and the value calculated using the approximate equation.

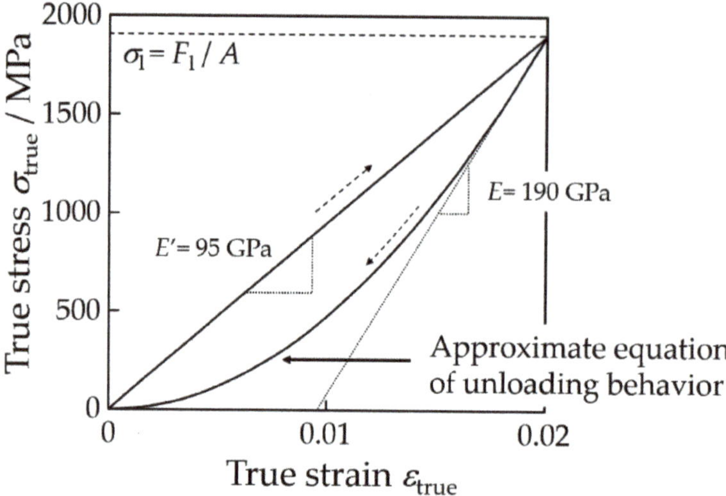

Figure 24. Unloading behavior when the true strain completely recovers. The parameter σ_1 is the drawing stress. The parameters F_1 and A are the drawing force and the cross-sectional area of the drawn tube, respectively. The parameters E and E' are the elastic modulus of the bulk metal and apparent elastic modulus, respectively.

4.2.4. Summary of Excessive Thinning of Outer Diameter

Figure 25 illustrates the excessive thinning of the outer diameter in the hollow sinking of a micro metal tube. The final outer diameter becomes smaller than the die diameter even when the drawing stress is in the elastic region and at any Lankford value. Furthermore, the outer diameter becomes smaller than the die diameter with an increase in the Lankford value, as shown by a theoretical formula (6). Therefore, it is considered that the micro tube yields microscopically during drawing, even under macroscopically elastic deformation behavior. The outer diameter seems to approach the die diameter during unloading more than the linear elastic strain outside the microscopic yielding region. Therefore, the dislocations generated during drawing are considered to disappear partially during unloading, which is similar to the Bauchinger effect. The unloading strain increases under the condition where few dislocations are generated during drawing, such as low drawing stress or high apparent elastic modulus of the stress-strain curve. Therefore, the final excessive thinning of the outer diameter decreases as the drawing stress decreases, or the apparent elastic modulus increases.

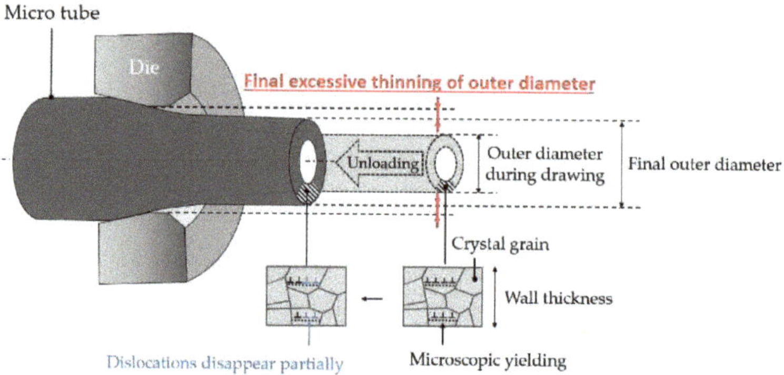

Figure 25. Schematic illustration of excessive thinning of the outer diameter during drawing and after unloading.

According to the above discussion, it is considered that the excessive thinning of the outer diameter can be suppressed by reducing the die half angle or the frictional force between the micro tube and the die, which results in reduced drawing stress for a particular yield stress. Furthermore, the drawing stress increases due to work hardening. However, since few dislocations are generated during the drawing of the micro tube, the drawing stress is much smaller than the yield stress because of the small size of the tube. Therefore, it is considered that work hardening of the micro tube suppresses the excessive thinning of the outer diameter.

5. Conclusions

The deformation behavior causing the excessive thinning of the outer diameter in the hollow sinking of micro thin-walled tubes can be explained as follows.

(1) At any Lankford value, the micro thin-walled tube yields microscopically during drawing even under a drawing stress lower than the yield stress of the bulk metal. Therefore, the outer diameter always becomes smaller than the die diameter during drawing.

(2) The outer diameter approaches the die diameter due to the unloading strain, which is larger than the linear elastic recovery, during unloading. The unloading strain increases as the drawing stress decreases or the apparent elastic modulus of the stress-strain curve increases. Therefore, the final excessive thinning of the outer diameter decreases as the drawing stress decreases, or the apparent elastic modulus increases. From the approximated unloading behavior, it is considered that the final outer diameter matches the die diameter completely during unloading due to elastic recovery when the apparent elastic modulus is larger than a threshold value.

Author Contributions: Conceptualization, T.K., K.T., and S.S. (Shinsuke Suzuki). Methodology, T.K., K.T., and S.S. (Shinsuke Suzuki). Validation, T.K., H.S., S.S. (Saki Suematsu), and S.S. (Shinsuke Suzuki). Formal analysis, T.K. Investigation, T.K., H.S., S.S. (Saki Suematsu), K.T., S.K., S.G., and S.S. (Shinsuke Suzuki). Data curation, T.K., H.S., S.S. (Saki Suematsu), and S.S. (Shinsuke Suzuki). Writing—original draft preparation, T.K. Writing—review and editing, T.K., H.S., S.S. (Saki Suematsu), K.T., S.K., S.G., and S.S. (Shinsuke Suzuki). Visualization, T.K., H.S., S.S. (Saki Suematsu), and S.S. (Shinsuke Suzuki). Supervision, S.S. (Shinsuke Suzuki). Project administration, S.S. (Shinsuke Suzuki). Funding acquisition, T.K., H.S., S.S. (Saki Suematsu), and S.S. (Shinsuke Suzuki). All authors have read and agreed to the published version of the manuscript.

Funding: This study was supported by JXTG Nippon Oil & Energy Corporation. Grant number is B2R50Z004300.

Acknowledgments: The first author performed this study as a Research Assistant of the Kagami Memorial Research Institute of Material Science and Technology of Waseda University.

Conflicts of Interest: The authors declare no conflict of interest.

Nomenclature

D_0	Outer diameter of starting material (mm)
D_{die}	Die diameter (mm)
D_{total}	Outer diameter of tube during drawing (mm)
D_n	Final outer diameter of drawn tube (mm)
E	Elastic modulus of bulk metal (GPa)
E'	Apparent elastic modulus of tube (GPa)
F	Load (N)
F_{BT}	Back tension during drawing (N)
F_1	Drawing tension during drawing (N)
h	Height of outer uneven surface of tube (μm)
H	Height from an arbitrary position of the outer uneven surface of the tube (μm)
H_{ave}	Average value of h (μm)
l_0	Length of starting material (mm)
l_{die}	Length of tube after passing through the die approach (mm)
l_{total}	Length of tube during drawing (mm)

Nomenclature

l_n	Final length of tube (mm)
R_e	Die reduction (-)
r	Lankford value of the tube (-)
t_0	Wall thickness of the starting material (mm)
t_{die}	Wall thickness of tube after passing through the die approach (mm)
t_{total}	Wall thickness of tube during drawing (mm)
t_n	Final wall thickness of tube (mm)
V_n	Drawing speed on the die's exit side (mm/s)
V_{n-1}	Drawing speed on the die's entrance side (mm/s)
β	Drawing speed ratio (-)
$\Delta\varepsilon_e$	Elastic strain (-)
$\Delta\varepsilon_E$	Excessive elastic strain (-)
ε_p	Plastic strain (-)
ε_{total}	Total strain (-)
ε_{true}	True strain (-)
$\Delta\varepsilon_{unload}$	Unloading strain (-)
η	Final excessive thinning of the outer diameter (%)
η'	Excessive thinning of the outer diameter during drawing (%)
θ	Die half angle (°)
σ_1	Drawing stress (MPa)
σ_{rue}	True stress (MPa)

Appendix A

The wall thickness of the micro tube after passing through the die t_{die} and the Lankford value r of deformation during drawing were calculated as follows.

It was assumed that the length of the micro tube after passing through the die l_{die} (state (ii) in Figure 17) and the length during drawing l_{total} (state (iii) in Figure 17) were the same. The length of the micro tube l_{total} could be expressed as $\beta \cdot l_0$. Therefore, the parameter l_{die} also could be expressed as $\beta \cdot l_0$. Equation (A1) indicates that the volume of the micro tube is constant in state (i) and (ii) in Figure 17.

$$l_0 \cdot t_0 \left(D_0 - t_0\right) = l_{die} \cdot t_{die} \left(D_{die} - t_{die}\right) \tag{A1}$$

By replacing l_{die} by $\beta \cdot l_0$, Equation (A2) is obtained.

$$t_{die} = \frac{1}{2}\left(D_{die} + \sqrt{D_{die} - \frac{4t_0(d_0 - t_0)}{\beta}}\right) \tag{A2}$$

The Lankford value r of deformation during drawing was defined by Equation (A3). The change of the inner diameter, which was considered in Equation (2), was neglected to simplify the calculation.

$$r = \frac{\ln\left(\frac{D_{total}}{D_{die}}\right)}{\ln\left(\frac{t_{total}}{t_{die}}\right)} \tag{A3}$$

It was assumed that the calculated Lankford value did not change significantly even if the dimensions during drawing (state (iii) in Figure 17) and the final dimensions (state (iv) in Figure 17) were the same. Therefore, by converting D_{total} and t_{total} into D_n and t_n, respectively in Equation (A3), Equation (A4) is obtained.

$$r = \frac{\ln\left(\frac{D_n}{D_{die}}\right)}{\ln\left(\frac{t_n}{t_{die}}\right)} \tag{A4}$$

References

1. Favier, D.; Liu, Y.; Orgeas, L.; Sandel, A.; Debove, L.; Comte-Gaz, P. Influence of thermomechanical processing on the superelastic properties of a Ni-rich nitinol shape memory alloy. *Mater. Sci. Eng. A* **2006**, *429*, 130–136. [CrossRef]

2. Xiong, S.; Wang, Q.; Chen, Y. Study on electrical conductivity of single polyani-line microtube. *Mater. Lett.* **2007**, *61*, 2965–2968. [CrossRef]

3. Tsuchiya, K.; Jinnin, S.; Yamamoto, H.; Uetsuji, Y.; Nakayama, E. Design and development of a biocompatible painless microneedle by the iron sputtering deposition method. *Precis. Eng.* **2010**, *34*, 461–466. [CrossRef]

4. Yoshida, K.; Furuya, H. Mandrel drawing and plug drawing of shape-memory-alloy fine tubes used in catheters and stents. *J. Mater. Process. Technol.* **2004**, *153–154*, 145–150. [CrossRef]

5. Yoshida, K.; Watanabe, M.; Ishikawa, H. Drawing of Ni-Ti shape-memory-alloy fine tubes in medical tests. *J. Mater. Process. Technol.* **2001**, *118*, 251–255. [CrossRef]

6. Takemoto, K. Tension control technology in non-slip type wire drawing machine. *J. Jpn. Soc. Technol. Plast.* **2016**, *57*, 1122–1125. (In Japanese) [CrossRef]

7. Kishimoto, T.; Gondo, S.; Takemoto, K.; Tashima, K.; Kajino, S.; Suzuki, S. Conditions for wall thickness reduction in hollow sinking of SUS304 tubes with drawing speed control in entrance and exit sides of die. *J. Manuf. Sci. Eng.* **2019**, *141*, 111008. [CrossRef]

8. Werkhoven, R.J.; Sillekens, W.H.; van Lieshout, J.B.J.M. Processing aspects of magnesium alloy stent tube. *Magnes. Technol.* **2011**, 419–424. [CrossRef]

9. Halaczek, D. Analysis of manufacturing bimetallic tubes by the cold drawing process. *Arch. Metall. Mater.* **2016**, *61*, 241–248. [CrossRef]

10. Kishimoto, T.; Sakaguchi, H.; Suematsu, S.; Tashima, K.; Kajino, S.; Gondo, S.; Suzuki, S. Outer Diameter and Surface Quality of Micro Metal Tubes in Hollow Sinking. *Procedia Manuf.* **2020**, *47*, 217–223. [CrossRef]

11. Vollertsen, F.; Hu, Z.; Niehoff, H.S.; Theiler, C. State of the art in micro forming and investigation into deep drawing. *J. Mater. Process. Technol.* **2004**, *151*, 70–79. [CrossRef]

12. Armstrong, R.W. On size effects in polycrystal plasticity. *J. Mech. Phys. Solids* **1961**, *9*, 196–199. [CrossRef]

13. Wang, C.; Wang, C.; DebinShan, B.; Huang, G. Size effect on flow stress in uniaxial compression of pure nickel cylinders with a few grains across thickness. *Mater. Lett.* **2013**, *106*, 294–296. [CrossRef]

14. Engel, U.; Eckstein, R. Microforming-from basic research to its realization. *J. Mater. Process. Technol.* **2002**, *125–126*, 35–44. [CrossRef]

15. Keller, C.; Hug, E.; Retoux, R.; Feaugas, X. TEM study of dislocation patterns in near-surface and core regions of deformed nickel polycrystals with few grains across the cross section. *Mech. Mater.* **2010**, *42*, 44–54. [CrossRef]

16. Celentano, D.J.; Rosales, D.A.; Peña, J.A. Simulation and Experimental Validation of Tube Sinking Drawing Processes. *Mater. Manuf. Process.* **2011**, *26*, 770–780. [CrossRef]

17. Huh, J.; Huh, H.; Lee, C.S. Effect of strain rate on plastic anisotropy of advanced high strength steel sheets. *Int. J. Plast.* **2013**, *44*, 23–46. [CrossRef]

18. Kitamura, K.; Terano, M. Determination of local properties of plastic anisotropy in thick plate by small-cube compression test for precise simulation of plate forging. *Manuf. Technol.* **2013**, *63*, 293–296. [CrossRef]

19. Japanese Industrial Standards Committee. *JIS G 4305, Cold-Rolled Stainless Steel Plate, Sheet and Strip*; Japanese Standards Association: Tokyo, Japan, 2012.

20. Japanese Industrial Standards Committee. *JIS H 4080, Aluminium and Aluminium Alloy Extruded Tubes and Cold-Drawn Tubes*; Japanese Standards Association: Tokyo, Japan, 2015.

21. Japanese Industrial Standards Committee. *JIS H 3300, Copper and Copper Alloy Seamless Pipes and Tubes*; Japanese Standards Association: Tokyo, Japan, 2018.

22. Akiyama, M.; Matsui, K.; Terada, K. Analysis of Microscopic yielding behavior of carbon steel under macroscopic loading. *Tetsu-to-Hagane* **2005**, *91*, 803–808. [CrossRef]

23. Zhang, J.; Nylas, A.; Obst, B. New technique for measuring the dynamic Young's modulus between 295 and 6K. *Cryogenics* **1991**, *31*, 884–889. [CrossRef]

24. Jiang, C.P.; Chen, C.C. Grain Size Effect on the springback behavior of the microtube in the press bending process. *Mater. Manuf. Process.* **2012**, *27*, 512–518. [CrossRef]

25. Kuboki, T.; Nishida, K.; Sakai, T.; Murata, M. Effect of plug on levelling of residual stress in tube drawing. *J. Mater. Process. Technol.* **2008**, *204*, 162–168. [CrossRef]
26. Japanese Aluminium Association. Physical Property. In *Aluminim Handbook*; Japanese Aluminium Association: Tokyo, Japan, 2001; pp. 26–29.

MDPI
St. Alban-Anlage 66
4052 Basel
Switzerland
Tel. +41 61 683 77 34
Fax +41 61 302 89 18
www.mdpi.com

Metals Editorial Office
E-mail: metals@mdpi.com
www.mdpi.com/journal/metals

www.ingramcontent.com/pod-product-compliance
Lightning Source LLC
LaVergne TN
LVHW070949170726
843515LV00005B/953